SpringerBriefs in Applied Sciences and Technology

Forensic and Medical Bioinformatics

More information about this series at http://www.springer.com/series/11910

Ravi Bhramaramba · Akula Chandra Sekhar
Editors

Application of Computational Intelligence to Biology

Editors
Ravi Bhramaramba
GITAM University
Visakhapatnam, Andhra Pradesh
India

Akula Chandra Sekhar
Sanketika Institute of Technology
and Management
Visakhapatnam, Andhra Pradesh
India

ISSN 2191-530X ISSN 2191-5318 (electronic)
SpringerBriefs in Applied Sciences and Technology
ISSN 2196-8845 ISSN 2196-8853 (electronic)
SpringerBriefs in Forensic and Medical Bioinformatics
ISBN 978-981-10-0390-5 ISBN 978-981-10-0391-2 (eBook)
DOI 10.1007/978-981-10-0391-2

Library of Congress Control Number: 2016937354

Printed on acid-free paper

This Springer imprint is published by Springer Nature
The registered company is Springer Science+Business Media Singapore Pte Ltd.

Preface

This book is about application of computational intelligence to biology.

Chapter "Enhancing the Performance of Multi-parameter Patient Monitors by Homogeneous Kernel Maps" deals with enhancing the performance of multi-parameter patient monitors by homogeneous kernel maps. Chapter "Augmenting the Performance of Multi-patient Parameter Monitoring System in LKSVM" deals with augmenting the performance of multi-parameter patient monitoring system in LKSVM. Chapter "An Efficient Classification Model Based on Ensemble of Fuzzy-Rough Classifier for Analysis of Medical Data" deals with an efficient classification model based on ensemble of fuzzy-rough classifier for analysis of medical data. Chapter "A Comparative Study of Various Minutiae Extraction Methods for Fingerprint Recognition Based on Score-Level Fusion" shows a comparative study of various minutiae extraction methods for fingerprint recognition based on score-level fusion. Chapter "Hybrid Model for Analysis of Abnormalities in Diabetic Cardiomyopathy" is about a hybrid model for the analysis of abnormalities in diabetic cardiomyopathy. Chapter "Computational Screening of DrugBank Database for Novel Cell Cycle Inhibitors" is concerned with computational screening of DrugBank database for novel cell cycle inhibitors. Chapter "Pathway Analysis of Highly Conserved Mitogen-Activated Protein Kinases (MAPKs)" is about pathway analysis of highly conserved mitogen-activated protein kinases. Chapter "Identification of Drug Targets from Integrated Database of Diabetes Mellitus Genes using Protein–Protein Interactions" focuses on identification of drug targets from integrated database of diabetes mellitus genes using protein–protein interactions. Chapter "Distributed Data Mining for Modeling and Prediction of Skin Condiction in Cosmetic Industry—A Rough Set Theory Approach" is based on distributed data mining for modeling and prediction of skin condition in cosmetic industry using a rough set theory approach.

Acknowledgments

At the outset, I thank the GITAM University authorities for permitting me to be the editor of this book. Next, I would like to thank Prof. Allam Appa Rao Sir for giving me this opportunity. I would also thank Dr. Amit Kumar for going through the papers and for having offered valuable suggestions. I am grateful to my fellow counterparts Dr. A. Chandra Sekhar, Dr. P. Satheesh, and Dr. A. Naresh. Lastly, I thank the Springer Press for accepting to publish this book.

Contents

Enhancing the Performance of Multi-parameter Patient Monitors by Homogeneous Kernel Maps

S. Premanand and S. Sugunavathy

Abstract A multi-parameter patient monitor (MPM) is a crucial appliance utilized in life-threatening care units of hospitals. It smells out a patient's vital signs without the demand of continuous attendance by the nurses. The operation of the SVM algorithm closely depends on the kernel mapping and their corresponding parameters. MPM stays informed regarding the state of a patient utilizing the baseline vital parameters, heart rate (HR), blood pressure (NIBP), respiration rate (RR) and oxygen saturation (SPO2). A high exactness for sensitivity, specificity and overall classification is critical in giving quality social insurance to the patients. Support Vector Machine (SVM) is an influential classification technique that is successfully used in the improvement of MPMs. A cautious investigation of the baseline parameters uncovers that they are constantly positive, making them suitable to be used with a class of kernels, such as intersection, Chi-Squared, and Jenson-Shannon kernels that are previously used effectively with histograms of images for scene classification. In our baseline system, we have utilized 10 features (four vital parameters along with six correlated features) based on our earlier research using decision tree algorithms. Using the intersection kernel, we obtained an improvement of 2.14 % in sensitivity, 1.38 % in specificity, and 1.57 % in the overall classification accuracy, over the best baseline system using radial basis function (RBF) kernel. In this work, we experiment with the histogram kernels and showed that intersection kernel can effectively improve the sensitivity, specificity, and overall classification accuracy of MPMs.

Keywords Correlation features · Homogeneous kernels · Multi-parameter patient monitor · Support vector machine and vital parameters

---

S. Premanand (✉)
Department of Electronics and Communication Engineering,
Amrita School of Engineering, Ettimadai, Coimbatore, Tamil Nadu, India
e-mail: newgen.anand@gmail.com

S. Sugunavathy
Department of Electronics and Communication Engineering,
Dr. Mahalingalam College of Engineering and Technology, Pollachi, Tamil Nadu, India

R. Bhramaramba and A.C. Sekhar (eds.), *Application of Computational Intelligence to Biology*, Springer Briefs in Forensic and Medical Bioinformatics,
DOI 10.1007/978-981-10-0391-2_1

1 Introduction

Successful, effective, and safe conveyance of wellbeing consideration is reliant on the convenient acknowledgment and treatment of a breaking down patient condition. Inability to protect patients in the previous phase of physiological crumbling can bring about lasting organ damage, expanded healing treatment, and expanded recovery time. MPM [1] are comprehensively used as the serious consideration as in intensive care units (ICU) and also in general wards to reliably screen patients' health wellbeing in human vital parameters like, HR, NIBP, RR and SPO2.

Investigation of different survey demonstrates the significance of vital parameters on physiological deterioration [2], (i) Patients regularly had a drawn out of times of physiological unsteadiness preceding admission to ICU, (ii) There were impressive time deferrals between physiological instability and successive critical care referral, (iii) Vital sign perceptions were once in a while accomplished for wiped out patients, even in the quick period before ICU admissions and (iv) The most frequently recorded vital signs were HR, NIBP, temperature, however RR is the slightest recorded variable, in spite of the way this is an early and delicate marker of decay. In the event that the physiological deterioration of the vital parameters is recognized at an early stage, the patient mortality can be progressed.

Early Warning Scores (EWS) [3] system in which, Patients' vital parameters are routinely recorded. With the EWS system each vital parameter is assigned a numerical score from 0 to 3, on a color coded perception outline (A score of 0 is more attractive and a score of 3 is minimum alluring). These scores are included and an aggregate score is recorded which arrives EWS [4]. A pattern can be seen whether the patients' condition is enhancing, with a bringing down of scores or dis-enhancing, with an increment of the score. Since the threshold value is fixed heuristically, the current EWS systems are not very precise. Further, they don't consider the characteristic relationship between the diverse vital parameters that exist in a healthy (ordinary) person [1, 5–7], and there is a noteworthy blunder rate connected with the manual scoring. When comparing to EWS, machine learning algorithms can be used effectively with high accuracy and as well as in cost reduction of patient health care. The data collected from the bedside monitors are used for training with some machine learning algorithm, and then used for predicting the condition of a new patient by using the model created using labelled data.

MPMs using SVM classifiers [4, 8] have been studied in the literature and found to be very effective in providing state-of-art performance. One class SVMs [4] performs novelty detection based on the prior intelligence gathered from some assumed "normal" behaviour. From the knowledge of the normal behaviour, it detects cases with abnormal to trigger alarms. All the same, if enough examples of patient worsening cases are available, a two-class SVM approach [8] may be approached. Due to more information in the two class SVM approach when compared to one class classifier, two class SVM outperforms experimentally [9].

Studies on physiological parameters [5–7] show that there exists a well-established intrinsic relationship between different vital parameters in a healthy person. For example, when the NIBP is high, HR is expected to be opposite of the NIBP i.e., low. Based on our previous research results [1], we further explore different avenues regarding the utilization of six correlation features, added to the baseline vital parameters making the aggregate number of features to ten, with an end goal to improve the execution of the MPM. The additional six correlation features denoted as correlation features catch the inborn relationship between the vital parameters and it was seen from the analyses that the utilization of correlation features upgraded the execution of the MPM altogether over the baseline MPM utilizing vital parameters.

$$A = [x1, x2, x3, x4, \sqrt{x1x2}, \sqrt{x1x3}, \sqrt{x1x4}, \sqrt{x2x3}, \sqrt{x2x4}, \sqrt{x3x4}]$$

where x1 meant for HR, x2 meant for NIBP, x3 meant for RR and x4 meant for SPO2 are the four vital parameters. Subsequently, we investigate the MPM along with correlation features, with intersection [11, 13], Chi Squared [12, 13] ad Jenson-Shannon (JS) [12, 13] kernels, and note that the performance of the MPM could be enhanced significantly with the use of intersection kernel over the best baseline system using RBF kernel.

2 Support Vector Machines

SVM [10, 15, 16] have a place with a class of supervised learning calculations in which the learning machine gives an arrangement of cases (classes) with the related labels. SVM develops a hyperplane that isolates the two cases, and the isolating hyper-plane is decided to amplify the edge between the samples of the two cases. The focuses in the dataset falling on the jumping (bounding) planes are called support vectors, *sv*. In the event that the information focuses are not straight (linearly) divisible, a non-linear kernel, k (x, y), may be utilized, where x and y are the two occasions in the dataset with their particular labels, lx and ly, with lx, ly $\in \pm 1$. Taking everything into account, the categorization of the test vector x may be characterized similar:

$$l_x = sign(\sum_{i=1}^{M} \alpha_i k(x, sv(i)))$$

where *sv(i)* speaks to the ith support vector, M is the quantity of support vectors, and α can be thought to be the double representation of the isolating hyperplane's ordinary vector.

In this study, we first try different things with three of the most prevalently utilized kernels [14, 15], linear, polynomial, and RBF kernels, in the SVM back-end utilized as a part of the MPM:

$$K_{lin}(x, y) = x^T \cdot y$$
$$K_{poly}(x, y) = (1 + x^T \cdot y)^d$$
$$K_{RBF}(x, y) = \exp(\frac{-||x - y||^2}{2\sigma^2})$$

The kernel, k (x, y), may be perused as a basis of similitude [14] between the two illustrations, x and y.

3 Proposed Method

SVM algorithm performance closely depends on the kernel function and their corresponding parameters (vital parameters). The most crucial kernels in the SVM classifications: linear kernel in linear classification, and in case of non-linear classification polynomial, RBF and sigmoid kernels are used. In this paper, we choice the intersection function and other homogeneous kernel techniques. Linear kernel SVM have turned out to be more prevalent, particularly in real-time applications. Because of their dot-product form, linear kernel SVMs have the capacity to be transmuted into a minimized form by substituting summation in classification formula, prompting to tremendously low complexity in terms of both time and storage. Nevertheless, linear kernel SVMs generally yield more terrible classification performance than that obtained by nonlinear kernel SVMs. The polynomial kernel has more hyper parameters and whereas vital parameters used in this experiment not validated with sigmoid kernel, hence this kernel is not used for the experiment. Even though RBF kernel which produced the best performance in the classification, still it has some disadvantage in the practical issues like the choice of the function parameters values of [sigma] and [epsilon].

It might be watched that the baseline parameters are constantly optimistic (positive) quantities, thus the six correlated features got from these baseline parameters too. This makes the MPM with the SVM upheld to qualify the utilization of intersection [11–13], χ 2 [12, 13], and JS [12, 13] kernels:

$$K_{int}(x, y) \sum_{i=1}^{N} \min(x_i, y_i)$$
$$K_{Chi}(x, y) = \sum_{i=1}^{N} \frac{x_i \cdot y_i}{x_i + y_i}$$
$$K_{JS}(x, y) = \sum_{i=1}^{N} \frac{x_i}{2} \log_2 \frac{x_i + y_i}{x_i} + \frac{y_i}{2} \log_2 \frac{x_i + y_i}{y_i}$$

where N is feature vector's dimension. The intersection non-linear SVM kernel is also popularly known as a histogram kernel in the survey, since they were first practiced with the histograms of images for scene classification.

4 Experiments and Results

MIMIC II database comprises of four baseline parameters, HR, NIBP, RR and SPO2, from 413 patients, information from 12 patients were not usable, and the remaining information from 401 patients is utilized for the analyses as a part of this work. We split the patient's data of 401 patient's as: 14,54,010 samples of 300 patient's as training segment and 3,11,423 samples of 101 patient's as testing segment. Afterwards, the full training data sets and as well as testing data sets were mixed randomly by using random permutation method, a subset of 50,000 samples with corresponding labels was selected as the training data, and also a subset of 20,000 samples from the test data set was selected as the testing data for one set of experimental trail set. Likewise, the process is redone to get samples for seven independent trials. The size of the data was reduced just for computational considerations, to avoid dealing with large kernel matrices in solving the model parameters. LIBSVM toolbox is used for all our experiments for the classification purpose of SVM in MATLAB.

The experiments starts from the baseline SVM system. Table 1 explains about the performance of vital parameters with baseline system, Table 2 demonstrate the performance of MPM with correlation features. In the case of polynomial kernel, 3 was found to be the optimum value of the order, d, whereas, 0.70 to be the optimum value for RBF kernel. Results indicate that the RBF kernel gave the best performance, for both four and ten features of the MPM system. It may be noted that though correlation features did help improve the performance of the MPM with RBF kernel, it is relatively less compared with other kernels. This may be attributed to the mapping of the features into an intrinsic relationship between the four vital parameters implicitly. Further, Table 3 gives the performance of the MPM

Table 1 Results of performance of MPM using 4 vital parameters with baseline system

Kernels	Class.Acc	Sensitivity	Specificity
Linear	76.74	2.215	100
Polynomial	93.06	81.17	96.76
RBF	98.11	96.17	98.59

Table 2 Results of performance of MPM using 10 vital parameters with baseline system

Kernels	Class.Acc	Sensitivity	Specificity
Linear	91.85	77.32	96.37
Polynomial	95.20	86.30	97.97
RBF	98.11	96.56	98.60

Table 3 Results of performance of MPM using 10 vital parameters with baseline system

Kernels	Class.Acc	Sensitivity	Specificity
Intersection	99.68	98.70	99.98
Chi square	93.95	82.31	97.57
JS	92.71	78.70	97.08

using ten features with the intersection, Chi-square and JS kernel gave the better performance results than baseline systems, and in that intersection kernel is better by 2.14 % in sensitivity, 1.38 % in specificity, and 1.57 % in the overall classification accuracy compared to the MPM using the RBF kernel with ten features.

5 Conclusions

We take note of that the estimations of the vital parameters, HR, NIBP, RR and SPO2 are constantly optimistic (positive)amounts, as are the correlation features got from these baseline vital parameters. We further watch that this property of the features makes them meet all requirements for the utilization of intersection, chi-square and Jenson-Shannon kernels when utilized as a part of MPMs with a SVM backend classifier. In our investigations with the intersection kernel for MPM, we had the capacity to bring a change of 2.14 % in sensitivity, 1.38 % in specificity, and 1.57 % in the classifications overall accuracy contrasted with the best standard framework utilizing the RBF kernel. We utilized ten features (four baseline parameters along with six correlated features) in both the framework.

References

1. Vaijeyanthi V, Vishnuprasad K, Santhosh Kumar C, Ramachandran KI, Gopinath R, Anand Kumar A, Yadav PK (2006) Towards enhancing the performance of multi-parameter patient monitors. Health 20(7):1483–1510
2. National Patient Safety Association, et al (2007) Safer care for acutely ill patients: learning from serious accidents. Tech Rep (NPSA)
3. Tarassenko L, Clifton DA, Pinsky MR, Hravnak MT, Woods JR, Watkinson PJ (2011) Centile-based early warning scores derived from statistical distributions of vital signs. Resuscitation 82(8):1013–1018
4. Clifton L, Clifton D, Watkinson PJ, Tarassenko L (2011) Identification of patient deterioration in vital-sign data using one-class support vector machines'. In: Computer science and information systems (FedCSIS), IEEE federated conference, pp 125–131
5. Beata G, Anna S, Krzysztof C, Wierlawa K, Grzegorz G et al (2013) Relationship between heart rate variability, blood pressure and arterial wall properties during air and oxygen breathing in healthy subjects. Auton Neurosci 178(1–2):60–66
6. Yasuma F, Hayano J (2004) Respiratory sinus arrhythmia: why does the heart beat synchronize with respiratory rhythm? Chest 125(2):683–690
7. Dornhost AC, Howard P, Leathart GL (1952) Respiratory variations in Blood pressure. Circulation 6:553–558

8. Khalid S, Clifton DA, Clifton L, Tarassenko L (2012) A Two-class approach to the detection of physiological deterioration in patient vital signs, with clinical label refinement. IEEE Trans Inf Technol Biomed 16(6):1231–1238
9. Bishop M (2006) Pattern recognition and machine learning. Springer, New York
10. Vapnik VN (1999) The nature of statistical learning theory, 2nd edn. Springer, Berlin, pp 237–240, 263–265, 291–299
11. Barla A, Odone F, Verri A (2003) Histogram intersection kernel for image classification. In: Proceedings of ICIP
12. Hein M, Bosquet O (2005) Hilbertian metrics and positive definite kernels on probability measures. In: Proceedings of AISTAT
13. Vedaldi A, Zisserman A (2011) Efficient additive kernels via explicit feature maps. In: IEEE Transactions on Pattern Analysis and Machine Intelligence, June 2011
14. www.physionet.org/physiobank/database/mimic2db/,13 June 2000
15. Scholkopf B, Smola AJ (2001) Learning with kernels—support vector machines, regularization, optimization, and beyond. MIT Press, Cambridge
16. Soman KP, Loganathan R, Ajay V (2009) Machine learning with SVM and other kernel methods. Prentice Hall India Learning Private Ltd., New Delhi

Augmenting the Performance of Multi-patient Parameter Monitoring System in LKSVM

S. Premanand and S. Sugunavathy

Abstract To offer quality wellness care to patients, multi-parameter patient monitors (MPM) need a high accuracy for sensitivity, specificity, and overall classification. Nevertheless, it is likewise important to provide affordable healthcare by providing cheap MPMs using todays handheld computing and communication devices, and low complexity hardware. Support vector machine (SVM) is a vital classification process valuable for the improvement of MPMs for its high exactness and viability in foreseeing the status of patients. It is well known that non-linear kernel SVMs offer better performance, while the linear kernel SVM (LKSVM) are computationally very efficient. This makes the LKSVM particularly attractive for low cost implementations. In this paper, we demonstrate that mapping feature to a higher dimension using locality-constrained linear coding (LLC), added to that the framework by eluding the system reliant features using dimensionality reduction technique called principal component analysis (PCA) to make the framework durable, which improve the execution of MPMs using LKSVM. It was seen that the use of LLC-PCA has helped enhance the sensitivity by 3.27 % from the baseline system.

Keywords Multi-parameter patient monitor · Support vector machine · Locality-constrained linear coding · Principal component analysis

S. Premanand (✉)
Department of Electronics and Communication Engineering, Amrita School of Engineering, Ettimadai, Coimbatore, Tamil Nadu, India
e-mail: newgen.anand@gmail.com

S. Sugunavathy
Department of Electronics and Communication Engineering, Dr. Mahalingam College of Engineering and Technology, Pollachi, Tamil Nadu, India

R. Bhramaramba and A.C. Sekhar (eds.), *Application of Computational Intelligence to Biology*, Springer Briefs in Forensic and Medical Bioinformatics,
DOI 10.1007/978-981-10-0391-2_2

1 Introduction

MPMs [1] make utilization of the indispensable signs of humans like, respiration rate (RR), heart rate (HR), blood pressure (BP) and oxygen saturation (SPO2) for observing the status of patients in escalated care units and inpatient wards. The utilization of this system in intensive care units (ICU) can recognize, identify the weakness in the patients' well being condition and start opportune intercessions to save lives. For the dependable execution of the MPMs, the likelihood of missing cautions (alarms) and in additional false alerts ought to be least, which implies the alert precision(sensitivity) and no-caution (alarm) accuracy (specificity) of the frame work ought to be similarly prominent as could be allowed.

Machine learning (ML) procedures were broadly utilized for the detecting of indispensable disintegration in patients' wellbeing. SVM is an effective classification technique, strategy which has been utilized for cataloguing purposes. SVM is all around perceived for its speculation capacity and effectiveness of arrangement even with higher measurement (dimensional) information. The proficiency of the framework execution, profoundly relies on upon the instance of the part being utilized as a part of the SVM. At the point when managing with linearly non-detachable information, non-linear kernel SVM (NLKSVM) give improved execution at the cost of expanded computational intricacy and capacity necessity.

One class SVMs [2], we model the "normal" condition of the patient using the normative data, and any deviation from the normal behavioral pattern of the vital parameter is a novelty or alarm condition. The performance of the MPM system could be improved using two class SVMs [3], when sufficient examples from the patient deteriorations are available

On the occasion of the monitor, the investigation of the indispensable parameters demonstrates that they are not directly (linear) separate. This affiliation is measured in SVMs by utilizing kernels that measure the closeness between any two cases. In the event that the samples fit in with same class the similitude ought to be augmented and in the event that they go to diverse classes, the closeness ought to be negligible. In this way, the effectiveness of the SVM classifier essentially relies on upon the decision of a fitting part of the kernel. Kernels capacities are assessed by deciding the similitude between data (information) focuses. On the account of linear kernels, the dot product between the two chosen data points focuses is the measure of closeness though in RBF (radial basis function) kernel, the opposite exponential of the Euclidian separation between the two is the measure of comparability.

In case of image classification, SVMs utilizing spatial pyramid matching [4] (SPM) part has been exceedingly effective especially in NLKSVM case. In SPM, a codebook with M sections is connected to quantize every vector and produces the higher dimensional coded vector C relating to the data vector Y that has a much lower measurement. In the event that hard vector quantization (VQ) is utilized, every code C has one and only non-zero component, while for delicate VQ, a little gathering of components can be nonzero. With a big number of images in the preparation information, for complex order issues, the amount of support vectors could be a couple of yards, and the computational prerequisites can be wholly critical.

Jianchao et al. [5, 6] Proposed sparse coded SPM (ScSPM) supplanting the VQ in the SPM with sparse coding (SC). The upside of the ScSPM is that the NLKSVM backend classifier in the SPM could be supplanted by a LKSVM backend classifier, without bargaining on the execution. The LKSVM lessens the preparation training, quality, and a consistent complexity in testing. SC productively speaks to the information as an in lines combination of an over complete premise set (codebook), in which the quantity of premise is more than that of data measurement of the feature vector [5, 6]. SC's quantization error is a great deal not exactly VQ coding as SPM.

Local coordinate coding (LCC) [7] is an adjustment of SC, which expressly urges the coding to be nearby. Kai et al. [7] clarified hypothetically, that under specific suppositions, locality is more vital than sparsity, as the locality prompts sparsity, while the opposite is not genuine. One primary detriment of the LCC is that it wants to get care of L1-norm optimization issue that is computationally lavish. Wang et al. [8] offered LLC which can be seen as a quick usage of LCC that uses the locality requirement to extend every vector into its neighborhood coordinate framework. What's more, the enhancement issue utilized by LLC has a scientific arrangement. To improve the computational many sided quality further, an approximated LLC technique is proposed in [8] by performing closest neighbor's pursuit to discover the K closest neighbor cells of the VQ centroids to the info vector, and afterward understanding an obliged minimum square fitting issue to discover the code to speak the data vector in the advanced dimensional space.

2 Locality-Constrained Linear Coding

Locality-constrained linear coding means a plotting technique (mapping) used to speak to non-linear features to a higher dimensional space to create them linearly distinct [9]. It has been broadly utilized in the application of image classification for its robust execution [8–10]. LLC technique strides can be explained in Fig. 1.

Where X meant for D-dimensional input vectors, B meant for codebook

$$X = [x1, x2, \ldots, xN] \in \mathbb{R}^{DXN}$$

$$B = [b1, b2, \ldots, bM] \in \mathbb{R}^{DXM}$$

$$\min_{C} \sum_{i=1}^{N} ||x_i - Bc_i||^2 + \lambda ||d_i \odot c_i||^2$$

$$s.t\ 1^T c_i = 1, \forall i$$

$$d_i = \exp\frac{dist(x_i, B)}{\sigma}$$

$$dist(x_i, B) = [dist(x_i, b_1), \ldots\ dist(x_i, b_M)]^T$$

$$\min_{C} \sum_{i=1}^{N} ||X_i - \tilde{c}_i B_i||^2$$

$$s.t.,\ 1^T \check{c}_i = 1, \forall \mathrm{i}$$

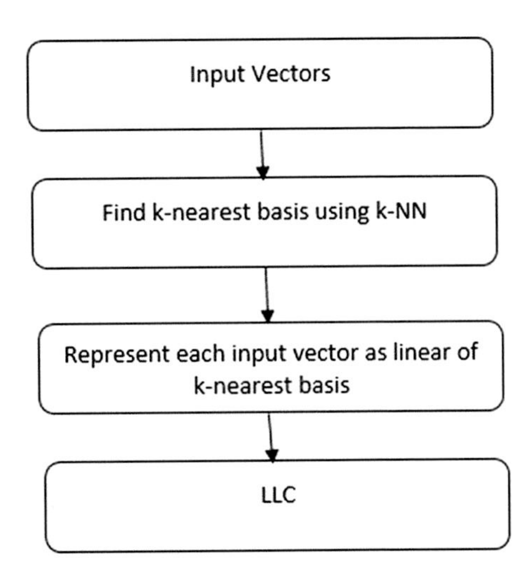

Fig. 1 Steps in LLC

LLC by fast approximation method lessens the computational complication. At that point we explore PCA with LLC to distinguish the framework-free features for augmenting the execution of the MPMs. We utilize LKSVM as a rear-end classifier.

3 Principal Component Analysis

Principal component analysis (PCA) [11–13] is a component change system used to tell apart the contours in the advanced dimensional information and waypoints the similitudes and divergences in the data (Fig. 2).

It is likewise used to diminish the quantity of measurements from the vectors, short of trailing the information. The strides included in the PCA are depicted, Fig. 3. In this work, we utilize PCA for distinguishing the framework-autonomous features over the framework.

4 LLC and PCA Method for Robust LKSVM

We take a note of that, with the assistance of LLC, we linearized the features into an advanced dimensional space to make them linearly isolated. Despite the fact LLC enhances the specificity exactness which brings about expansion in general classification precision, affectability precision was not enhanced because of framework reliant information. In this manner, we have to identify and uproot the framework-reliant features to enhance the execution of MPMs. The strategy of LLC-PCA framework for the vigorous MPM framework is appeared in the Fig. 4.

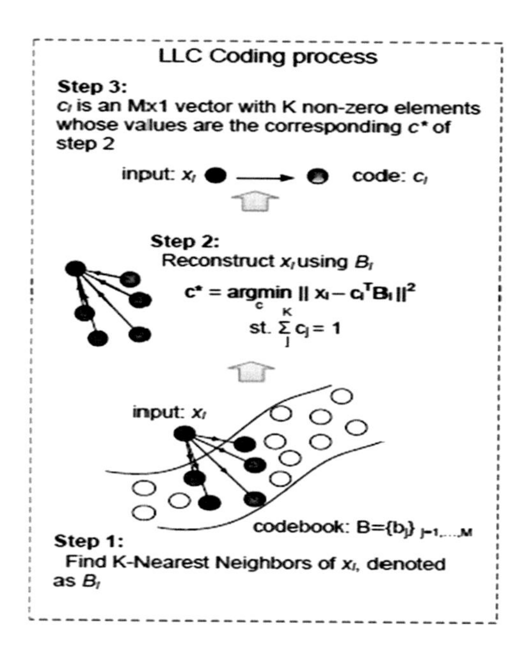

Fig. 2 LLC coding technique

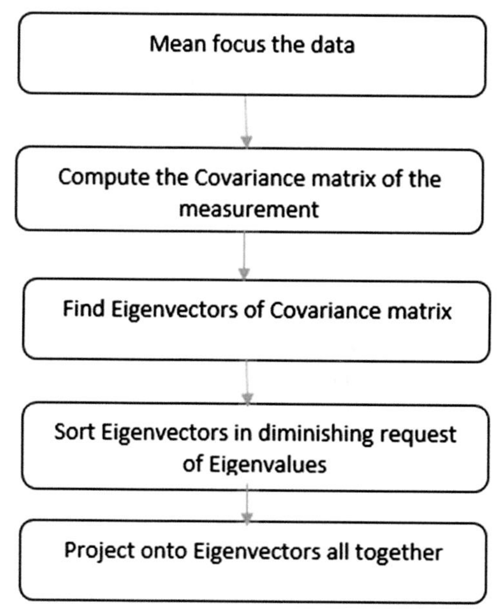

Fig. 3 Steps in principle component analysis

From our examinations with this technique, we similarly base, few features don't have valuable data subsequent to mapping to a sophisticated dimensional space.

SVMs [14, 15] have produced as a prevalent means to deal with ML, for grouping and regression approach, exhibiting condition of-craftsmanship execution in differing applications and offering an alluring distinct option for counterfeit neural

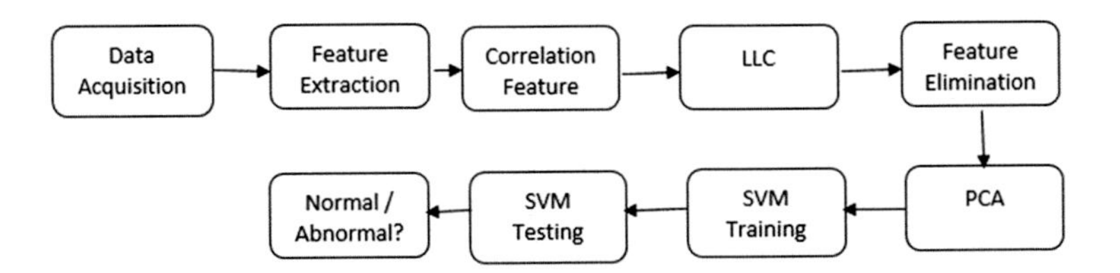

Fig. 4 Proposed methodology support vector machine

system and master-based methodologies. It builds a hyperplane that isolates two groups indicated in Fig. 5. At the same time, the SVM approach tries to accomplish most extreme partition between the categories. Isolating the classes with an extensive edge minimizes a bound on the normal speculation mistake. A 'minimum generalization error', implies that when new instances touch the base for classification, the likeliness of getting a wrong confidence in the forecast in view of educated classifier ought to be unimportant.

Vapnik and Vapnik [16] has demonstrated that if the preparation vectors (train section) are isolated without mistakes by an ideal hyperplane, the normal blunder rate on a test is restricted by the proportion of the desire of the support vectors (SV) to the quantity of the preparing vectors. Since this proportion is liberated of the measurement of the issue, if one can acquire a humble arrangement of SV, great speculation is ensured. Hyperplane boosts the edge and it can be acquired by deciding the separation between bounding planes to the cause separately (b1, b2) and subtracting the separation (b2–b1), to augment the edge, SVM can be planned as minimization issue and communicated just as equally,

$$\frac{2}{||w||^2} \rightarrow \min_{x,\gamma} \frac{||w||^2}{2}$$

Subject to the constraint:

$$d_i\left[\left(w^T \cdot x_i\right)\right] + \gamma \geq 1$$

Fig. 5 LKSVM process

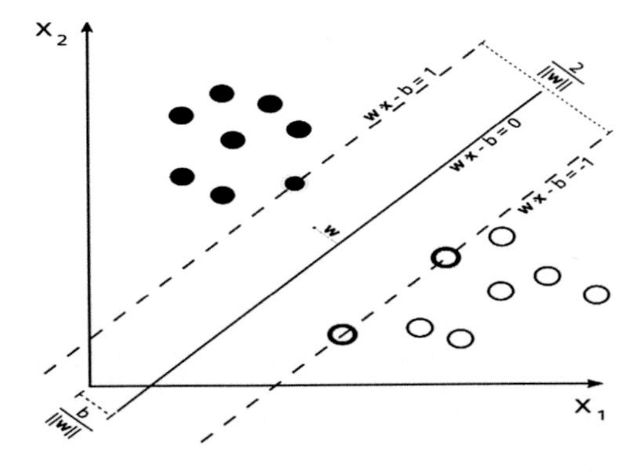

The ideal arrangement can be acquired by Lagrangian dual technique, and SVM learning formulation will be,

$$\max_{\alpha \geq 0} \min_{w,\gamma} \left\{ \frac{1}{2}||w||^2 - \sum_{i=1}^{n} \alpha_i \left[d_i \left(w^T \cdot x_i - \gamma \right) - 1 \right] \right\}$$

α_i denotes Lagrangian Multiplier. The issue can be illuminated as, by using KKT condition (Karush-Kuhn-Tucker)

$$\max_{\alpha_i} \left\{ \sum_{i=1}^{n} \alpha_i - \frac{1}{2} \sum_{i,j} \alpha_i \alpha_j \, d_i d_j k\left(x_i, x_j\right) \right\}$$

Subject to:

$$\alpha_i \geq 0, \sum_{i=1}^{n} \alpha_i \, d_i = 0$$

The decision function can be expressed equally, γ can be computed by compelling a point on the edge for which $\alpha_i \neq 0$.

$$f(x) = sign\left(w^T x - \gamma\right) = sign(\sum_{i=1}^{n} \alpha_i d_i x_i^T x - \gamma)$$

5 Experiments and Results

We utilized MIMIC-II databank [17, 18] for every one of our tests in this study. MIMIC II databank contains baseline parameters gathered from 413 patients. The database also comprises of a screen produced cautions alongside the doctor explained documents. Among the information from 413 patients, we chose the information from 401 patients for this test based on the tone of the information from the patient. We split the information from 401 patients, as 14,54,010 samples into training from 300 patients and 11,00,510 samples as test sets from 101 patients by using a random permutation method. Subsequently, the entire training data were shuffled randomly and a subset of 50,000 samples with their corresponding labels was selected as the training data for one trial of the experiment, and a subset of 20,000 samples from the test set was selected as the test data for the corresponding trial. Another set of training and test data is selected from the remaining data, to generate the data for another trial. The method is iterated to get tested for seven free trials, and the answers acquired from these seven tests are found the middle value to get the last reply. The size of the data was reduced just for computational considerations, to avoid dealing with large kernel matrices in solving the model parameters. We used LIBSVM toolbox [19] for all our experiments with SVM.

Foremost, we developed an MPM consuming the baseline parameters. Based on previous research results [20], we then used correlation features, geometric mean of two vital parameters required at a time, in addition to the four vital parameters, causing the total number of features to ten. It was determined that the role of correlation features helped enhance the overall classification accuracy to 14.8 %. We preferred this system as our baseline system for comparing the potency of utilizing an LLC to map the features to a higher dimension to improve the operation of the MPM, using the LKSVM backend classifier.

To cause the LLC coding, we first generated a VQ codebook using Linde-Buzo-Gray (LBG) [21, 22] clustering algorithm till about 2048 clusters, only the training information was employed for generating the codebook. We experiment with different k-NN and codebook sizes for better classification performance. In our experiments, we note that for 1024 clusters was found to be the optimum value for the codebook size (M), 19 for the number of nearest neighbors, K.

In this work, features are linearized into advanced dimensional space by utilizing LLC and that has no information with it are wiped out, henceforth it improves no alarm (specificity) condition which enhances overall accuracy for classifier but in the case of sensitivity condition the system is not enhanced because of reliable features. Thus we go for dimensionality reduction technique PCA to recognize and get rid of the framework reliant-features. We tested LLC-PCA technique for distinctive codebook sizes and k-closest neighbors for more reliant working. We exactly found, for 1024 cluster codebook size, with a choice of 180 dimensions using dimensionality technique, the execution has been enhanced when contrasted with the standard framework and LLC for the strong MPM model. For a system should be in robust condition and should be reliable for all kind of classification it not only depends on overall classification but also satisfies both sensitivity and specificity condition also.

Table 1 compares the performance of different MPM systems. It may be noted that the sensitivity increased by 3.27 %, specificity of 0.46 %, and the overall classification accuracy of 0.35 % with the use of an LLC-PCA methodology. We enhanced the affectability precision (sensitivity) for the MPM system framework model to the detriment of specificity exactness. It might be noticed that a superior affectability precision is coveted in basic social insurance applications even to the detriment of a lower specificity exactness.

Table 1 Comparability of results for enhanced LKSVM for different features

Input data	D	OA	SEN	SPE
Vital parameter	4	77.14	1.55	100
Vital + Corr.	10	91.94	77.09	96.43
With LLC	1024	97.67	92.57	99.21
LLC-PCA	180	98.02	95.84	99.67

D Dimension of the feature
OA Overall Accuracy
SEN Sensitivity
SPE Specificity

6 Conclusions

It is really important to deliver high sensitivity, specificity, and overall classification accuracy, for MPM to provide quality health care. However, it is also important to provide affordable healthcare, by being able to make the MPM system algorithm using low cost computing and communication devices, and low complexity hardware.

It is surely understood that LKSVM is well known for its computational proficiency, and its suitability to ease execution. In this paper, we explored the use of the LLC technique to linearize the component vectors to a higher dimensional space and after that we chose the framework free elements utilizing dimensional reduction technique PCA to upgrade the execution of the MPM with a LKSVM backend classifier.

References

1. National Patient Safety Association (2007) Safer care for acutely ill patients: learning from serious accidents. Tech Rep (NPSA)
2. Clifton L, Clifton, DA, Watkinson PJ, Tarassenko L (2011) Identification of patient deterioration in vital-sign data using one-class support vector machines. In: Federated conference on computer science and information systems (Fed-CSIS) (Szczecin, Poland). IEEE, pp 125–131
3. Khalid S, Clifton DA, Clifton L, Tarassenko L (2012) A two-class approach to the detection of physiological deterioration in patient vital signs, with clinical label refinement. In: IEEE transactions on information technology in biomedicine, vol 16, no 6, pp 1231–1238
4. Lazebnik S, Schmid C, Ponce J (2006) Beyond bags of features: spatial pyramid matching for recognizing natural scene categories. In: IEEE computer society conference on computer vision and pattern recognition
5. Jianchao Y, Yu K, Gong Y, Huang T (2009) Linear spatial pyramid matching using sparse coding for image classification. In: IEEE Conference on Computer Vision and Pattern Recognition (CVPR), Miami, FL, USA, pp 1794–1801
6. Lee H, Battle A, Raina R, Ng AY (2007) Efficient sparse coding algorithms. In: Advances in neural information processing systems, vol 19, p 801
7. Kai Y, Zhang T, Gong Y (2009) Nonlinear learning using local coordinate coding. NIPS 9:1
8. Wang J, Yang J, Yu K, Lv F, Huang T, Gong Y (2010) Locality-constrained linear coding for image classification. In: IEEE conference on computer vision and pattern recognition (CVPR), pp 3360–3367
9. Taniguchi K, Han XH, Iwamoto Y, Sasatani S, Chen YW (2012) Image super-resolution based on locality-constrained linear coding. In: 21st IEEE international conference on pattern recognition (ICPR), pp 1948–1951
10. Zhang P, Wee CY, Niethammer M, Shen D, Yap PT (2013) Large deformation image classification using generalized locality-constrained linear coding. In: Medical image computing and computer-assisted intervention (MICCAI 2013), pp 292–299
11. Hyvärinen L (1970) Principal component analysis, mathematical modeling for industrial processes. Springer, Berlin, pp 82–104
12. Moore B (1981) Principle component analysis in linear systems: controllability, observability, and model reduction. In: Automatic Control, IEEE Transactions, pp 17–32, Feb 1981
13. Kim KI (2002) Face recognition using kernel principle component analysis. IEEE Sig Process Lett 9(2):40–42

14. Soman KP, Loganathan R, Ajay V (2009) Machine learning with SVM and other kernel methods. Prentice Hall India Learning Private Ltd., New Delhi
15. Scholkopf B, Smola AJ (2001) Learningwith kernels—support vector machines, regularization, optimization, and beyond. MIT Press, Cambridge
16. Vapnik VN, Vapnik V (1998) Statistical learning theory, vol 2. Wiley, New York
17. Goldberger AL, Amaral LAN, Glass L, Hausdorff JM, Ivanov PCH, Mark RG, Mietus JE, Moody GB, Peng CK, Stanley HE (2000) Physio Bank, Physio Toolkit and PhysioNet: Components of a new research resource for complex physiologic signals. Circulation 101(23):215–220
18. Lee J, Scott D, Villarroel M, Clifford G, Saeed M, Mark R (2011) Open-access MIMIC-II database for intensive care research. In: 33rd annual international conference of the IEEE EMBS. Boston, Massachusetts, USA, pp 8315–8318, Sept 2011
19. Chang C-C, Lin C (2011) LIBSVM: A library for support vector machines. ACM T Intell Syst Technol 2(3):27:1–27:27
20. Tarassenko L, Clifton DA, Pinsky MR, Hravnak MT, Woods JR, Watkinson PJ (2011) Centile-based early warning scores derived from statistical distributions of vital signs. Resuscitation 82(8):1013–1018
21. Gray RM (1984) Vector quantization. IEEE ASSP Magazine, pp 4–29, Apr 1984
22. Linde Y, Buzo A, Gray RM (1980) An algorithm for vector quantizer design. In: IEEE Transactions on Communications, pp 702–710, Jan 1980

An Efficient Classification Model Based on Ensemble of Fuzzy-Rough Classifier for Analysis of Medical Data

M. Sujatha, G. Lavanya Devi, N. Naresh and K. Srinivasa Rao

Abstract Clinical databases have been accumulated with large amounts of data due to the advances in the medical field. A medical dataset usually contains objects/records of patients that include a set of symptoms that a patient experiences. Medical data will have uncertainty due to the reason that a patient suffering from a specific illness cannot be completely determined by one or more symptoms; a certain set of symptoms can only indicate that there is a probability of a particular illness. Analysis of such medical data could reveal new insights that would definitely help in efficient diagnosis and also in drug discovery. This paper proposes a fuzzy-rough set based rule induction classifier to analyze medical data. In addition, we have presented a rough set based data preprocessing approach.

Keywords Medical data · Uncertainty · Data preprocessing · Rough set theory · Fuzzy-rough classifier

1 Introduction

The advents which occurred in the domain of both hardware and software in the past decade has immersed the world in the ocean of data. On the other hand, scientists around the globe are constantly working to develop new techniques for extracting knowledge from the data. However, these techniques are to be designed according to the domain of the data. Though a very good number of methods have been proposed for analyzing data in data mining, there is always a scope and need for new methodologies to address the challenges being faced by the heterogeneous, redundant and high dimensional data.

M. Sujatha (✉) · G. Lavanya Devi · N. Naresh · K. Srinivasa Rao
Department of Computer Science and Systems Engineering, AU College of Engineering, Andhra University, Visakhapatnam, Andhra Pradesh, India
e-mail: sujathamadugulacse@gmail.com

R. Bhramaramba and A.C. Sekhar (eds.), *Application of Computational Intelligence to Biology*, Springer Briefs in Forensic and Medical Bioinformatics,
DOI 10.1007/978-981-10-0391-2_3

A medical dataset usually contains objects/records of patients that include a set of symptoms that a patient experiences. Medical data will have uncertainty due to the reason that a patient suffering from a specific illness cannot be completely determined by one or more symptoms; a certain set of symptoms can only indicate that there is a probability of a particular illness.

Analysis of such medical data could reveal new insights that would definitely help in efficient diagnosis and also in drug discovery. Uncertainty in medical data is due to lack of precise knowledge or insufficient information. This scenario in the real world leads to uncertainty in the diagnosis of illness. The data mining [1] methods are inefficient in dealing with cognitive uncertainties such as vagueness and ambiguity. In general, Vagueness occurs due to insufficient data about some elements of the universe. Ambiguity is identified with two or more preferences such that the choice among them is left undefined.

Medical data also suffers with curse of dimensionality. High dimensionality (number of features) could include redundant or irrelevant features that have a noisy effect on the data mining technique. Rough set theory (RST) is used in medical field to address vagueness in terms of exact/crisp concepts known as *lower approximation* and *upper approximation*. RST is also used to identify the optimal reducts (attributes) in a given data set. However, a limitation of RST is in its inability to handle continuous features in the dataset. Nevertheless, this limitation can be improved using discretization techniques. But this generally comes with a cost of information loss. The need to handle both crisp and continuous features in the given dataset can be provided by using fuzzy-rough set theory (FRST). Thus, RST can be extended to FRST to handle the dissimilar impression of vagueness and indiscernibility, which arises uncertainty in data.

Active research going on in this area yielded a few data mining techniques based on fuzzy set theory (FST) and RST. These techniques in general identify relationships between data elements, classify the data and perform prediction [2] based on the patterns identified.

In this paper, a fuzzy-rough classification model for medical datasets is proposed. In addition, a novel method for data preprocessing based on RST is adopted. A quick reduct algorithm based on RST for identifying best attributes is presented. A rule-based classification method (classifier) based on FRST [3] is built on reducts. Finally, to enhance the prediction accuracy the proposed classifier is incorporated as a base classifier in the bagging based ensembler to perform classification.

The rest of this paper is structured as follows. Overview of the proposed system is presented in Sect. 2. Section 3 describes the proposed rough set based data preprocessing methods for medical dataset. Section 4 presents an ensemble of fuzzy-rough rule based classifier for medical datasets. Section 5 attribute selection based on RST to find optimal reducts. Finally, future directions of this work are presented in conclusion.

2 Overview of the Proposed System

The proposed system accepts medical datasets as input and predicts the class label of the given instance with the aid of the learning model developed. The developed system comprises of Data preprocessing, Base Classifier Pool Generation, Fuzzy-Rough Classifier and a Decision Aggregation Pool. A medical data set is represented as a table called as decision system. First, the primary operations of RST are performed to compute indiscernibility relation, lower and upper approximations and positive region for the given decision system. Then, data preprocessing for missing value imputation and instance selection using the aforementioned RST concepts is performed.

RSMVI algorithm applied to fill the missing values by using operations of RST. Selection of quality objects/instances is determined by the Fuzzy-Rough Instance Selection (Fuzzy-Rough IS) Algorithm is developed by using the information in positive region. The classification model is built on bagging based ensemble of fuzzy-rough classifier. In this model, the learning from the decision system is done through reducts that cover the decision system to generate optimal fuzzy rules defining knowledge as IF-THEN statements (See Fig. 1).

However, it is observed that in such rules antecedent are covering almost all attributes of the decision system which eventually increase the computational time in classification. To overcome this issue we propose to combine the rule induction and attribute selection process. Thereby the fuzzy rules are generated from the best

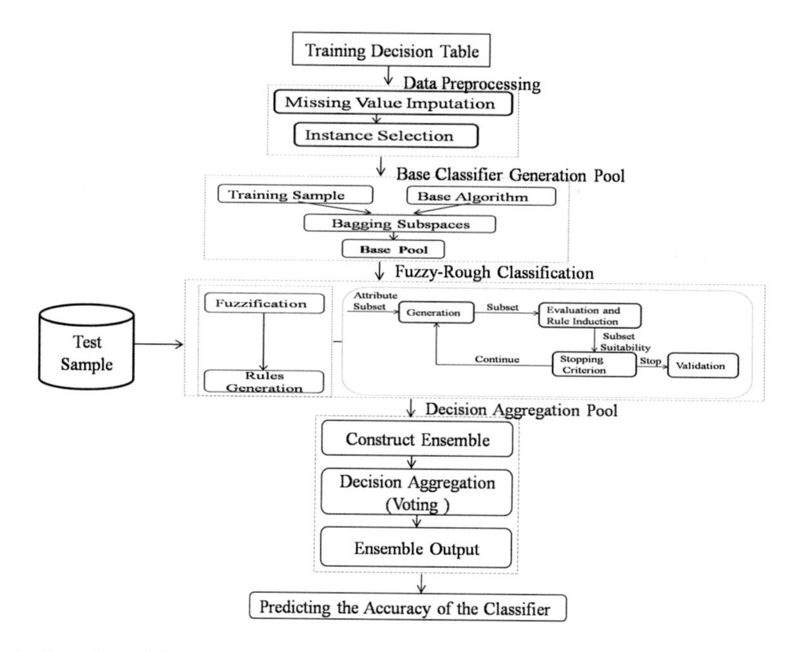

Fig. 1 Overview of the proposed system

attributes that maximally cover the decision system. Attribute selection process is done by *Rough Set Based Quick Reduct Algorithm* (RSQRA), which employs RST concepts to generate the best attribute subset in the decision system by computing dependency of attributes subset which equals to dependency of the given decision system. The following sections furnish more details of the methods used and also give the algorithms designed for the proposed system.

3 Rough Set Based Data Preprocessing

Often, data in real-world is redundant, conflicting and also contains numerous errors. These issues can be resolved by performing data preprocessing. Data preprocessing reduces the impurities in raw data for further processing. Data is made to go through cleansing phase where the imputations of missing data, softening the noisy data or analyzing the inconsistencies in the data are performed.

Missing value is a common occurrence in datasets and can have a significant effect on the conclusions that can be drawn from the data. The missing data imputation is a challenging task in data cleaning. In this work, we proposed a rough set based missing value imputation for accurate prediction of lacking values. And also when the size of the dataset is large, the data mining techniques cannot be definitely finalized without reducing the size of the data. This challenge can be overcome by the selection of potential instances removing the redundant and irrelevant instances.

3.1 Rough Set Based Missing Value Imputation Approach (RSMVI)

The presence of lost/missing data in the medical dataset can effect of the classification accuracy. Many imputation methods are proposed to substitution/replacing of missing values by plausible values. However, their usage in the domain of data mining is limited. Here, we propose an imputation technique depend on RST to measure and substitute missing data in a given medical dataset. The major RST operations needed in the proposed method are

- A given dataset is represented in terms of an information system I.
- Indiscernibility relation (*IND*) is computed to identify a relation between two objects or more, where all the values are identical in relation to a subset of conditional attributes A.
- The lower approximation is computed to determine the relationships between conditional attributes A and decision attributes D and specify which objects *definitely* belong to the concept $X \subseteq U$.

- The upper approximation is computed to determine the relationships between conditional attributes and decision attributes and specify which objects *possibly* to the concept $X \subseteq U$.

RSMVI Algorithm

Input Given medical dataset with missing values incomplete data set as information system I with conditional attributes A and d is decision values. All these decision values should belong to a decision D.
Output a vector containing possible missing values.
Assumption decision attributes D and some conditional attributes A will always be known.

(a) A given dataset is represented in terms of an information system $I = (U, A \cup D)$, where U is a finite, non-empty set of objects called the universe of discourse, A is a finite, non-empty set of attributes, such that a: $U \rightarrow V_a$ for every $a \in A$, where V_a is the set of values that the attribute a may take and $D \notin A$ is a decision attribute.
(b) Indiscernibility relation (*IND*) operation is performed to partition the universe of objects U according to decision attributes D into equivalence classes.
(c) For each conditional attribute A, repeat the following steps:
(d) For all attributes compute the family of equivalence classes $[x]_p$.
(e) The degree of belongingness γ is calculated to extract relationships between set of objects U. If a_i has the similar conditional attribute values with a_j apart from the missing value, replace the missing value, a_{missig}, with the value v_j, from a_j, where j is an index to another instance.
(f) Otherwise compute the lower approximation of each conditional attribute A, given the available data of the same instance with the missing value.
(g) While performing this process, if more than one v_j value is suitable for the estimation, delay the replacement till the value is suitable.
(h) Else, compute the upper approximations of each subset partition. Perform the computation and imputation of missing data as done with the lower approximation. Either *crisp* sets will be found, otherwise, *rough* sets can be used and missing data can be heuristically be selected from the obtained rough set.
(i) Enumerate the upper approximations of every subset division. Implementing the missing value imputation technique in 6. On the other hand *crisp* sets or *rough* sets are generated. The lost data can be examined and chosen from RST.

3.2 *Fuzzy-Rough Instance Selection (Fuzzy-Rough IS)*

The Fuzzy-Rough IS eliminates or substitutes redundant, noisy or uncertain instances from medical datasets but preserve consistent ones for future. In RST aspect, it classifies every object covered in the boundary region. Particularly, the

Fig. 2 Fuzzy-rough IS algorithm

$$
\begin{array}{l}
\textit{Fuzzy}-\textit{Rough Instance Selection}\ (S,\alpha,\beta).\\
\quad S, \textit{the set of object to be reduced};\\
\quad\quad \alpha, \textit{the granularity parameter};\\
\quad\quad\quad \beta, \textit{a selection threshold}.\\
\quad\quad\quad\quad Y \leftarrow S\\
\quad\quad\quad\quad \textbf{foreach}\ x \in S\\
\quad\quad\quad\quad\quad\quad \textbf{if}\ (POS_{a}^{\alpha,S}(x) < \beta)\\
\quad\quad\quad\quad\quad\quad\quad\quad Y \leftarrow Y - \{x\}\\
\quad\quad\quad\quad \textbf{Return}\ Y
\end{array}
$$

objects in lower approximations are protected, however the objects covered in the boundary region are refined, for illustration eliminating them or modifying their class labels to be persistent.

The proposed Fuzzy-Rough IS algorithm requires an input of three parameters, where first parameter is the set of objects, second parameter is the fuzzy similarity measure alpha α, and the last parameter is an arbitrary parameter β for discarding a greater number of records/instances. Generally, β set to 1.

The Fuzzy-Rough IS algorithm estimates the degree of membership of every object x to the positive region ($\mathrm{POS}_{A}^{\alpha\,s}(x)$); if positive region ($\mathrm{POS}_{A}^{\alpha\,s}(x)$) value is below the threshold value, then the objects might be eliminated. Meanwhile an object membership is below one, this situation indicates a bit of uncertainty so that classify this objects truly belongs. If all such objects are discarded, then there is no unlikeness presented by the left over objects Y (See Fig. 2).

4 Ensemble of Fuzzy-Rough Classifier

The primary objective of an ensemble classifier [4] is to get better performance over a single classifier. Due to the diversity in the internal models of various classifiers, their predictions [5] are not always the same on the certain datasets. By exploiting the errors which are uncorrelated, the accuracy of the classifier can be improved by combining different classifiers. An individual ensemble member are trained on a subset of training data, which to some extent decreases the computational complexity issue than to a single classifier is applied to large datasets. A common approach for constructing classifier ensembles involves building a cluster of classifiers by bagging technique to produce the final prediction.

4.1 Fuzzy-Rough Rule Based Induction

The RST is successfully applied to discover hidden examples/patterns in the information system and also for generating decision rules. However, a limitation for RST is its inability to handle continuous data. This limitation is overcome by constructing a model combining fuzzy set and RST [6]. The lower and upper approximations of RST are fuzzified by adopting by following approach.

- The set A may be generalized to a fuzzy set in X, allowing that objects can belong to a given concept (i.e., meet its characteristics) to varying degrees.
- Rather than assessing objects' indiscernibility, we may measure their approximate equality, represented by a fuzzy relation R. As a result, objects are categorized into classes, or granules, with "soft" boundaries based on their similarity to one another. As such, abrupt transitions between classes are replaced by gradual ones, allowing that an element can belong (to varying degrees) to more than one class. We present an algorithm for classification of the medical datasets hybridizing fuzzy and rough set theory.

Fuzzy Rough Rule based Induction Algorithm

From the given medical dataset M, select randomly a subset P of conditional attribute.

i. Initially a subset P of conditional attribute A is empty; an empty rule set R and cover set Cov is empty.
ii. For each attribute $a \in M$, repeat the following steps.
    - For each object $o_1 \in M$, repeat the following steps.
    - Compute degree $D1$ of belongingness of o_1 to positive region of attribute a.
    - Compute degree D2 of belongingness of o_1 to positive region for given dataset M.

iii. If degree D1 equals to D2

Construct the rule r, for the object o_1 and attribute subset $P \cup a$.
Add the rule to rule set R if r does not have same or more coverage then existing rules in rule set R. and also update the coverage set.

This algorithm generates fuzzy-rough If-then rules for prediction on given medical dataset.

However, it is observed that in such rules antecedent are covering almost all attributes of the decision system which eventually increase the computational time in classification. To overcome this issue we propose to combine the rule induction and attribute selection process. Thereby the fuzzy rules are generated from the best attributes that maximally cover the decision system. Attribute selection process is done by *Rough Set Based Quick Reduct Algorithm*.

4.2 RST Based Attribute Subset Selection

RST is a tool to find data dependencies and to select the attributes from dataset that are most informative [7–9]. A RST based attribute selection algorithm using wrapper approach is presented in this paper. The proposed algorithm analyzes the

Fig. 3 RSQR algorithm

Input: C, th set of all conditional features; D the set of decision features
output: R, the feature subset
$R \leftarrow \{\}$
while $\gamma_R(D) \neq \gamma_C(D)$
return R
foreach $x \in (C - R)$
if $\gamma_{R\cup\{x\}}(D) > \gamma_T(D)$
$T \leftarrow \gamma_{R\cup\{x\}}$
$R \leftarrow T$

degree of dependency of attributes using indiscernibility relation, the fundamental operations of RST [10] and identifies the best reducts of the given dataset.

RSQR Algorithm

The RSQR algorithm finds reducts without exhaustively making all available subsets. Initially, RSQR algorithm starts with an empty set reduct R. For each conditional attribute(C) of the given information system (I) the algorithm computes its degree of dependency with decision attribute (D).

The RSQR Algorithm(C, D) (See Fig. 3).

The degree of dependencies is computed using the total number of subsets of lower, upper approximations of the indiscernibility relations and total number of objects. Then, the attributes having highest dependency degree are expanded to set T. This technique progress until the dependencies of the attributes of reduct (R) set greater than or equals to the degree of the conditional attributes (C) dependences. Finally, the best reducts of the given data set are available in R.

5 Conclusion

This work articulates the power of FRST in combating uncertainty in the form of vagueness and ambiguity, often present in medical data. A fuzzy rough rule based classifier is developed to efficiently handle uncertainties in medical dataset, which can be an aid in effective diagnosis and also in drug discovery. A novel approach based on RST for imputation of missing values is presented. Fuzzy-rough instance selection method for identifying the potential instances from dataset to reduce the size of the data is presented. We have also incorporated attribute subset selection task in the rule induction process instead of data pre-processing phase. Some directions for future research include, the integration of RST with other intelligent constituents of soft computing paradigm such as neural networks, evolutionary computing for classification and rule generation for medical data. Swarm intelligence can be combined with RST to find the minimal reducts in the given data set.

References

1. Witten IH, Frank E (2005) Data mining: practical machine learning tools and techniques. Morgan Kaufmann
2. Hong J (2011) An improved prediction model based on fuzzy-rough set neural network. Int J Comput Theor Eng 3(1):1793–8
3. Little RJA, Rubin DB (1987) Statistical analysis with missing data. Wiley, New York
4. Maimon O, Rokach L (2005) Decomposition methodology for knowledge discovery and data mining: theory and applications. World Scientific, Singapore
5. Rokach L, Maimon O, Arad O (2005) Improving supervised learning by sample decomposition. Int J Comput Intell Appl 5(1):37–54
6. Dubois D, Prade H (1992) Putting rough sets and fuzzy sets together. In: Slowinski R (ed) Intelligent decision support: handbook of applications and advances of the rough sets theory. Kluwer Academic Publishers, Boston, pp 203–222
7. Pawlak Z (1991) Rough sets: theoretical aspects of reasoning about data. Kluwer Academic Publishing, Dordrecht
8. Polkowski L (2002) Rough sets: mathematical foundations. Advances in soft computing. Physica Verlag, Heidelberg, Germany
9. Zhao J, Zhang Z (2011) Fuzzy-rough neural network and its application to feature selection. Int J Fuzzy Syst 13(4)
10. Mac Parthalain N, Jensen R (2010) Measures for unsupervised fuzzy-rough feature selection. Int J Hybrid Intell Syst 7:1C11

A Comparative Study of Various Minutiae Extraction Methods for Fingerprint Recognition Based on Score Level Fusion

P. Aruna Kumari and G. JayaSuma

Abstract A Multimodal Biometric system combines the evidences from various biometric sources or multiple evidences from single biometric source to atone for the limitations in performance of unimodal biometric system. This paper discusses two Minutiae extraction techniques to recognize fingerprint based on confidence level fusion of two extracted features, bifurcations and ridge endings and compares the recognition accuracy. In particular, the well known Morphological based minutiae extraction approach is compared with the proposed fuzzy logic control based approach. Experimental results based on IITD fingerprint database demonstrate that the score level fusion of bifurcations and ridge endings for fingerprint leads to a dramatically improvement in performance. And also the results reveal that our proposed fuzzy logic control based minutiae extraction is much more reliable than the Morphological based minutiae extraction approach.

Keywords Multimodal biometrics · Fingerprint · Fuzzy logic control · Morphology · Score level fusion · ROC

1 Introduction

The up-growing advances in information systems and raise in security requirements have shown the way to reliable authentication of a person. In this context, because of the properties which cannot be stolen, borrowed, and forgotten, the

P. Aruna Kumari (✉)
Department of Computer Science and Engineering, JNTUK-UCEV,
Vizianagaram, Andhra Pradesh, India
e-mail: arunakumarip.cse@jntukucev.ac.in

G. JayaSuma
Department of Information Technology, JNTUK-UCEV, Vizianagaram,
Andhra Pradesh, India
e-mail: gjs.cse@gmail.com

R. Bhramaramba and A.C. Sekhar (eds.), *Application of Computational Intelligence to Biology*, Springer Briefs in Forensic and Medical Bioinformatics,
DOI 10.1007/978-981-10-0391-2_4

biometric based authentication has taken the leading role. Biometric authentication system makes use of physiological (e.g., face, fingerprint, palm print) and/or behavioural characteristics (e.g., voice, gait) of a person for the purpose of recognition [1]. It is more capable of discrimination between an authenticated or genuine person and imposter or fraudulent person than conventional knowledge or token based methods. Among various biometrics, fingerprint recognition is the most trustworthy and hopeful personal recognition technologies and which have been positioned in a wide range of applications, varying from forensics to mobiles [2].

Though, the performance of unimodal biometric system is affected by noise, sample size, and spoofing attacks [3], multimodal biometric systems can conquer a number of these problems by combining the features from a single trait or more than one biometric trait. In multimodal biometric systems, the evidences can be integrated at various levels: fusion at sensor level, feature level, score level, and decision level [4]. The fusion of evidences at feature level is hard because they may be incompatible or suffer from high dimensionality. The decision level fusion is too unbending because of the limited availability of information. Therefore, the fusion at score level is preferred due to simple in accessing and integration of scores [5]. In score level fusion, fusion can be performed in two distinct approaches namely classification approach and combined approach. [6, 7] shows that the combination approach performs better when compared with classification approach.

The flow of ridge patterns on the tip of the finger is known as a fingerprint [8] and whose pattern constitutes different minutiae points like ridge endings, trifurcations and bifurcations. Most of the researches selected bifurcation and ridge endings, which are prominent minutiae points in fingerprint recognition [9]. In the literature, a wide range of methods have been projected for the automatic minutiae extraction from fingerprint. Majority of the methods [10–12] extract the minutiae points from the thinned binary image of the fingerprint.

In most of researches, the extraction of minutiae from raw fingerprint image takes the following steps like preprocessing, edge detection, binarization and thinning. The advantage of binarization is it greatly reduces the amount of the data to be considered in thinning step. But, at the same time it has undesirable effect on some significant information loss [13]. Because the gray-values are useful in preserving the connectivity of ridge lines before thinning, two techniques are proposed in literature. In [14], a two-dimensional ridge detection algorithm is applied and [15] proposed a modified Deutch algorithm to extract furrow counters and ridges and then fuzzy based feature selection is applied.

A large number of false minutiae points can be extracted from a poor quality fingerprint image. Postprocessing is essential to remove those false minutiae, by that the recognition rate can be improved. Most of the researches eliminated false minutiae points based on some statistical evaluations [16], based on both structural and statistical information [17], and a bridge structured method [18].

1.1 Proposed Methodology

The architecture of the system is shown in Fig. 1. The significant points in the proposed methodology are:

- The crucial component of this paper is to examine the fuzzy logic for extracting minutiae points from fingerprint. The fuzzy logic has been utilized for biometric based authentication at classification level [19]. The literature shows no experiments have been carried in minutiae extraction level.
- The complexity of image data can be handled by Fuzzy logic and also output information is reduced. Moreover the vagueness present in a fingerprint image can be handled by fuzzy logic in an efficient way and for similar performance it needs less expensive hardware. In this work, fuzzy inference system is developed to extract bifurcation and ridge ending point using Crossing Number.
- The previous works in minutiae extraction are mainly based on morphological operations, which is compared with proposed method. In this, the image undergoes various preprocessing steps like enhancement, binarization, edge detection and then morphological operations have been applied. This image will be thinned and minutiae points marked.
- In the proposed method, after enhancing the image, fuzzy inference system is applied on the gray scale image then marks minutiae points.
- In Matching, the matching scores are integrated using weighed score level fusion. The fused score is compared against user specific threshold to identify genuine or imposter user.

The rest of paper is planned as follows: Sect. 2 presents steps involved in preprocessing and extracting minutiae from skeletonized images. A thorough study of fuzzy logic control based minutiae extraction method from gray scale images is presented in Sect. 3. False minutiae points are removed to increase the performance of recognition system, is discussed in Sect. 4. Matching and weighed score level fusion of the evidences from fingerprint image is given in Sect. 5. The experimental results and analysis are discussed in Sect. 6. Section 7 concludes this paper.

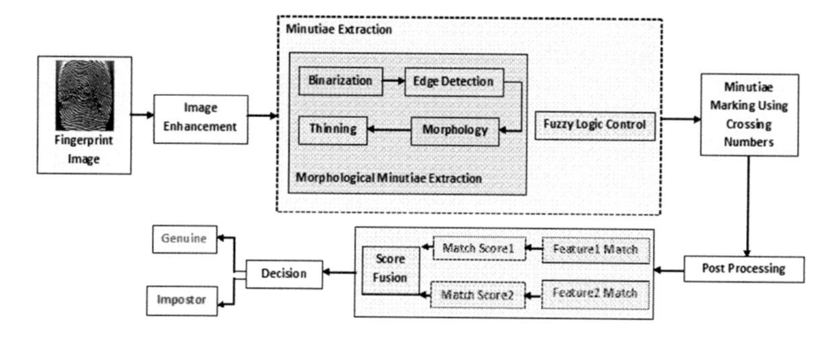

Fig. 1 Proposed methodology architecture

2 Preprocessing and Minutiae Extraction from Skeleton Images

A significant step in fingerprint recognition is extracting reliable minutiae points. From the given raw fingerprint image, extracting minutiae through its binary image is normally contains of the following steps.

2.1 *Image Enhancement and Binarization*

The performance of extracting minutiae is driven by the quality of the given fingerprint image. A huge number of false minutiae are extracted from low quality fingerprint image. Therefore, it is essential to enhance the raw fingerprint image for connecting false broken points of ridges before extracting minutiae. In [20–22], various methods have been proposed for improving the quality of fingerprint image. We have used the Histogram equalization for expanding the pixel value distribution of the fingerprint which increases the perceptional information.

Binarization transforms a grey scale image into a binary image which contains only two levels, black and white. The binarization in one way starts by selecting a threshold value, and then classifying every pixel with the value above this threshold as white and the rest of all pixels as black. Among the various methods proposed [2, 12, 23], this paper applied adaptive binarization where an optimal value of threshold is selected for each area of image, the images are shown in Fig. 2.

2.2 *Edge Detection Using Sobel Filter*

The majority of edge detection methods work on the assumption that the edge occurs where there is a discontinuity in the intensity function or a very steep intensity gradient in the image. Based on this assumption, the edge can be identified at which the derivative of the intensity value across the image is maximum, which is shown in Fig. 3. The components of a gradient vector measures the change in rapid pixel value with distance in the x_1 and y_1 direction. Thus, the gradient components can be found using the following calculation

$$\frac{\partial f(x_1,y_1)}{\partial x_1} = \Delta x_1 = \frac{f(x_1+dx_1,y_1) - f(x_1,y_1)}{dx_1} \tag{1}$$

$$\frac{\partial f(x_1,y_1)}{\partial x_1} = \Delta y_1 = \frac{f(x_1,y_1+dx_1) - f(x_1,y)}{dy_1} \tag{2}$$

Where dx_1 and dy_1 represents the distance along the x_1 and y_1 directions respectively. The distances dx_1 and dy_1 are considered as the pixel count between two points. The pixel point (p, q) where $dx_1 = dy_1 = 1$ thus,

$$\Delta x_1 = f(p+1, q) - f(p, q) \tag{3}$$

$$\Delta y_1 = f(p, q+1) - f(p, q) \tag{4}$$

The discontinuity can be detected by calculating the change in the gradient at (p, q). The direction θ of gradient is given by:

$$\theta = \tan^{-1} \frac{\Delta x_1}{\Delta y_1} \tag{5}$$

2.3 Morphological Operations

As mathematical morphology emphasizes in shape information, it finds a place in computer vision. As the analysis in morphology is based on lattice, set theory, random functions, and topology, in that sense it is mathematical [23]. By clean examination of the binarized fingerprint image, the isolated regions like dots, holes, islands and misconnections add a number of spurious minutiae. The morphological operations like dilation, erosion, opening and closing are applied to the binary image to eliminate these spurious minutiae.

Dilation grows or thickens object whereas Erosion thins or shrinks object in a binarized image. The pixels are added to the boundaries of objects in an image by Dilation. Erosion eliminates pixel to the boundaries of objects in an image. The spatial attributes of an image like contours, break narrow isthmuses are smoothened by opening. It eliminates false touching, thin ridges, small islands, and branches. On the other hand, closing smoothens the spatial attributes of an image like contours. It eliminates small holes and fills small gaps, narrow gulfs.

Fig. 2 Preprocessing of an image

2.4 Thinning

In further step, the morphological thinning operation is applied on the processed binary image. The thinning algorithm eliminates superfluous pixels from ridges until the ridges are just one pixel wide [24]. The thinning can be performed by using an iterative, parallel thinning algorithm. In each scan of the full fingerprint image, the algorithm identifies superfluous pixels in each small image window (3 × 3) and finally eliminates all those identified pixels after a number of scans.

2.5 Minutiae Extraction

The accuracy of minutiae extraction drives the reliability of any fingerprint recognition system. The Crossing Number (*CN*) concept is extensively used for minutiae extraction [10–12, 18]. In [25], Rutovitz's given the definition of crossing number for a pixel as

P_4	P_3	P_2
P_5	P	P_1
P_6	P_7	P_8

$$CN = 0.5 \sum_{i=1}^{8} |P_i - P_{i+1}|$$

where P_i is the neighborhood binary pixel value of P_i with $P = (0 \text{ or } 1)$ and $P_1 = P_9$.

Based on the properties of CN all minutiae points are marked from the thinned fingerprint image as shown in the Fig. 4. By considering the 3 × 3 window, if 1 is

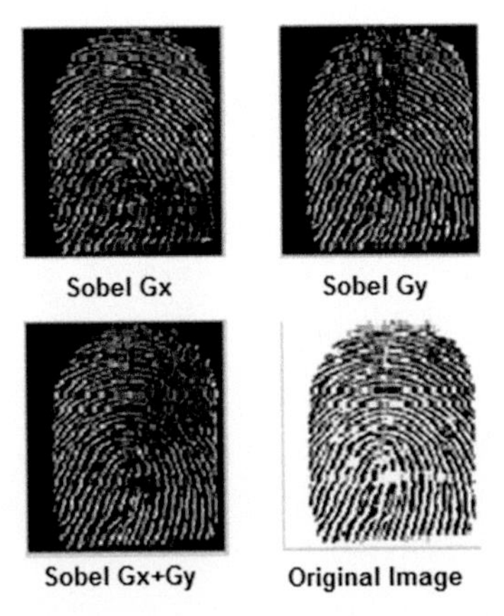

Fig. 3 Edge detection using sobel filter

Fig. 4 Crossing number properties

CN	Property
0	Isolated Point
1	Ending Point
2	Connective Point
3	Bifurcation Point
4	Crossing Point

0	1	0
0	1	0
1	0	1

Bifurcation Point (CN=3)

0	1	0
0	1	0
0	0	1

Ridge Ending Point (CN=2)

Fig. 5 Minutiae marking in morphology based method

the value in central pixel and has exactly three neighbor pixels with value 1 then it is identified as bifurcation point. If 1 is the value in central pixel and has exactly one neighbor pixel with value 1 then it is recognized as ridge ending. For morphology based feature extraction the minutiae marking is shown in Fig. 5. Though preprocessed the raw fingerprint image, both true and false minutiae are extracted. The post processing eliminates these false minutiae points.

3 Fuzzy Logic Control Based Minutiae Extraction

In a basic way, pattern recognition identifies or searches various patterns or structure in given data. The presence of vagueness or fuzziness in most of the real world entities leads to inaccurate results. This fact establishes the relation between the fuzzy logic and pattern recognition theories. And fuzzy logic facilitates the reasoning with the specific form of knowledge. In Biometric images, due to some sort of noise or for other reasons the vagueness is present at pixel level. Fuzzy logic can handle this type of vagueness at pixel level.

At the same time, the boundary between two uniform regions in any image is called an edge. These edges can be detected by comparing the luminance of neighboring pixels. However, the luminance at uniform region is not a crisp value and small variations among two neighboring pixels do not form a true edge. These variations form spurious minutiae in fingerprint. Here, fuzzy logic control based method can be applied by using any membership function to decide whether the pixel belongs to edge or uniform region. Because of edge connectivity preserving property of gray scale image, it is better to extract minutiae from gray image when compared to binary image. The thinned fingerprint image can be directly obtained by applying fuzzy logic control based method on gray scale image. To obtain this, the following steps are performed:

1. Convert raw finger print to gray scale image.
2. Since the gradient of an image identifies break points in uniform region, calculate the gradient of fingerprint along x-direction and y-direction. In literature, the gradient has been calculated by applying sobel filter, prewitt filter, simple gradient filter, and Gaussian filter. In this work, simple gradient filters G_x, G_y have been applied. Then the gray fingerprint image is convolved with G_x, G_y separately to get gradients along x-direction and y-direction respectively. The obtained gradient values will be in the rage $[-1, 1]$.
3. Once the gradients are obtained, edges can be detected by defining fuzzy inference system. In this for each gradient input Gaussian membership function is applied. Then for fuzzified input gradients triangular membership function has been applied to get output.
4. Then fuzzy inference system rules are defined to identify pixels. The following rules have been applied in this method:

Rule 1 If the pixel value along x-direction is zero and y-direction is zero then the pixel belongs to uniform region and it does not form any edge.

Rule 2 If the pixel value along x-direction is not zero or y-direction is not zero then the pixel belongs to an edge.

Since, crisp values are required, defuzzify the fuzzy values. Then minutiae points, ridge endings and bifurcations are marked based on Crossing Number. The false minutiae are removed in post processing to get true minutiae points.

4 Post Processing

The extracted minutiae points in the previous step involve many spurious or false minutiae points shown in Fig. 6. The algorithm initially finds the distances between termination and bifurcation. In this algorithm, the distances are calculated using Euclidian distance. Once distance is calculated, by writing some rules false minutiae points are removed.

Applied fuzzy rules are as follows:

Rule 1 If the distance between termination and bifurcation is less than a threshold T and they are in the same ridge, eliminate both of them (merge, bridge, ladder, lake).

Rule 2 If the difference between bifurcations orientation angles is very small and the distance between them is less than a threshold T and, then the two minutiae are considered as spurious (break, multiple breaks) and are eliminated.

5 Matching and Score Level Fusion

The Euclidean distances from the bifurcation points and ridge ending points are calculated then by using weighted sum approach those scores are combined. These approaches try to minimize the FRR for a given FAR. By using basic operations like sum, min, max, or product, the matching scores from various modalities can be integrated. The weights assigned for all evidences can be matcher specific, user specific or based on quality of sample. The weights assigned must be relative to the matching accuracy in matcher specific weighting. In user specific, the selection of weights is based on how much extent the matcher knows for a specific person. In the same way, the weights are allotted based on the quality of evidences given to matcher in quality based weighting.

5.1 *Fusion of Bifurcation and Ridge Endings Distance Scores*

The confidence level fusion based on sum rule is a very promising fusion method in multimodal biometrics. The basic form of integration takes the weighted average of the matching scores from the multiple evidences. Given the similarity scores of bifurcation and ridge ending, then the linear integration of two scores (fused score) is generated as,

$$S = \frac{(1-\alpha)S_{bifurcation} + \alpha S_{ridgeending}}{2} \tag{6}$$

6 Experimental Results

The experiments are carried out on IITD fingerprint database which consists of 20 objects and for each object 10 samples, total 200 samples are taken. The results shows that how much reliable the proposed fuzzy logic control based minutiae extraction when compared to morphological based one. Comparisons of various

Table 1 **a** For morphological based minutiae extraction-some of the threshold values and respective FAR and FRR values, **b** for fuzzy logic control based minutiae extraction-some of the threshold values and respective FAR and FRR values

a			b		
Threshold	FAR	FRR	Threshold	FAR	FRR
0.5	0.42	0.26	**0.5**	**0.22**	**0.26**
0.6	0.36	0.29	0.6	0.26	0.29
0.7	0.39	0.34	0.7	0.39	0.33
0.75	**0.29**	**0.33**	0.75	0.39	0.34
0.8	0.22	0.4	0.8	0.39	0.35
0.82	0.3	0.4	0.82	0.37	0.4
0.85	0.17	0.5	0.85	0.37	0.42
0.87	0.18	0.5	0.87	0.18	0.42
0.9	0	0.65	0.9	0.11	0.5
0.92	0	0.65	0.92	0.01	0.5

bold font indicates that the values where highest accuracy is attained

fingerprints of different individuals produce inter person values and intra person values are given by comparison of various samples of the same person.

The performance metrics in our analysis are False Acceptance Rate (FAR), False Rejection Rate (FAR). The acceptance rate of imposter is called FAR and the rejection rate of genuine person as an imposter is called FRR.

To calculate FAR, the evidences of every user are compared with other user's evidences in the database. While these comparisons, if the similarity score is greater than the user defined threshold value, then it indicates that a fraudulent is accepted. In every comparison of evidences from the different samples of single user, if the match score is more than the user defined threshold, then it indicates that the user is a genuine user and otherwise it indicates that an authentic user is not accepted i.e. erroneously rejected. This gives FRR the values obtained for the above metrics at various thresholds in morphological based minutiae extraction are shown in Table 1a. These results show that the best accuracy by morphology based minutiae extraction approach at confidence level fusion is 69 %. The values obtained for the fuzzy logic control based minutiae extraction are tabulated in Table 1b. At the same threshold values, the best accuracy produced by the proposed fuzzy based minutiae extraction at confidence level fusion is 76 % (Fig. 6).

ROC curve illustrates the measured accuracy of proposed fingerprint recognition system. ROC curves are generated by changing the user specific threshold on the categorization image and an optimal value of the threshold is selected at the maximum accuracy. The ROC curve, which depicts the correlation between FAR and FRR for morphological based approach is shown in Fig. 7a. The ROC curve for fuzzy logic control based approach is shown in Fig. 7b. The performance of fuzzy logic control based method shows that there is a remarkable increase in the recognition rate of multimodal system against the morphological based method. The comparison of accuracy rates of the minutiae extraction methods are shown in Fig. 8.

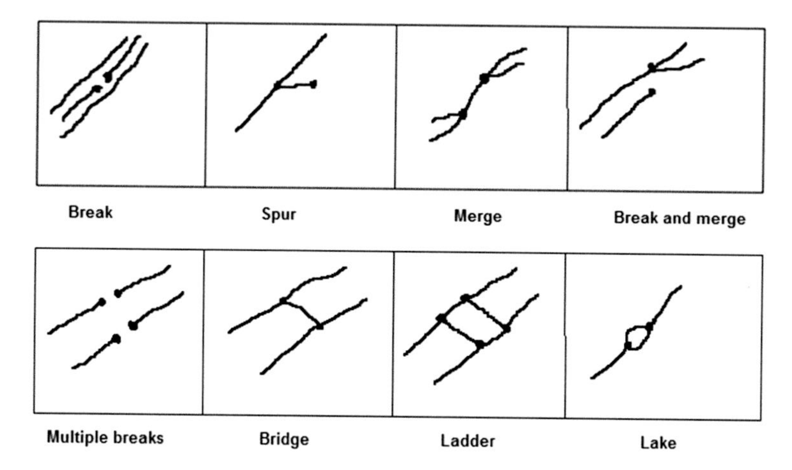

Fig. 6 Few common false minutiae points

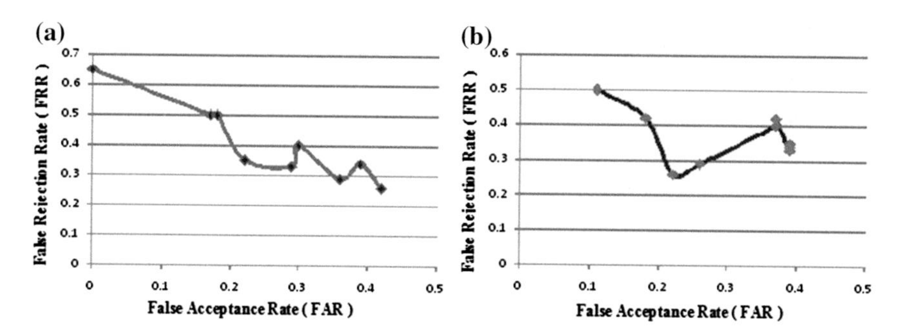

Fig. 7 **a** ROC curve between FAR and FRR for morphology based minutiae extraction, **b** ROC curve between FAR and FRR for fuzzy logic control based minutiae extraction

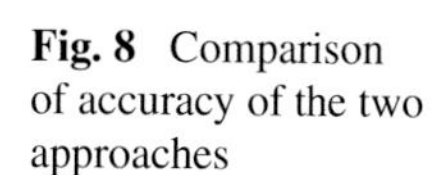
Fig. 8 Comparison of accuracy of the two approaches

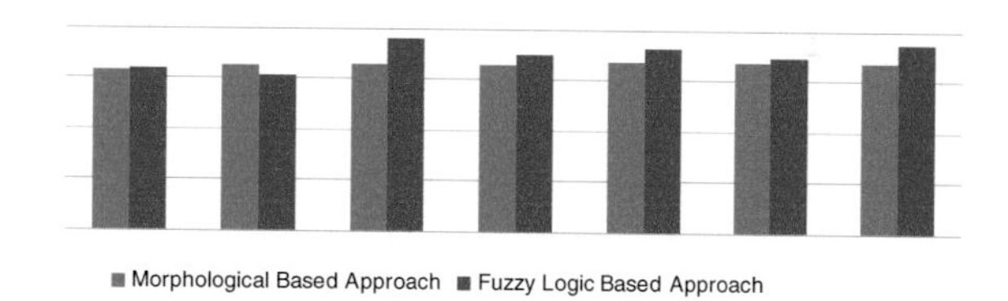

7 Conclusion

Two different minutiae extraction techniques for fingerprint recognition were investigated and presented the results on their accuracy. These experiments are carried out by considering two features from fingerprint and performed weighed score level fusion. The experimental results indicated that the proposed fuzzy logic control based minutiae extraction method have the well noticed improvement in accuracy. A drastic change in the accuracy was observed in proposed method. The Morphological based approach has 69 % of accuracy where as fuzzy logic control based approach has 76 % of accuracy in weighed score level fusion. In future work, our focus will be on improvement in accuracy. In this context, a classification based approach using k-NN and C4.5 classifiers are planned to examine at decision level.

References

1. Jain AK, Ross A, Prabhakar S (2004) An introduction to biometric recognition. IEEE Trans Circ Syst Video Technol 14(1):4–20 (special issue on image- and video-based biometric)
2. Maltoni D, Maio D, Jain AK, Prabhakar S (2009) Hand book of fingerprint recognition. Springer, Berlin
3. Cui FF, Yang GP (2011) Score level fusion of fingerprint and finger vein recognition. J Comput Inf Syst 7:5723–5731
4. Ross AA, Nandakumar K, Jain AK (2006) Handbook of multibiometrics. Springer, Berlin
5. Jain A et al (2005) Score Normalization in multimodal biometric systems. Pattern Recognit 38:2270–2285
6. Ross A, Jain AK (2003) Information fusion in biometrics. Pattern Recognit Lett 24(13):2115–2125 (special issue on multimodal biometrics)
7. Lip CC, Ramli DA (2012) Comparative study on feature, score and decision level fusion schemes for robust multibiometric systems. In: Sambath S, Zhu E (eds) Frontiers in computer education, AISC 133. Springer, Berlin, pp 941–948
8. Jain AK, Ross AA, Nandakumar K (2011) Introduction to biometrics. Springer, Berlin
9. Lu H, Jiang X, Yan W-Y (2002) Effective and efficient fingerprint image post processing, vol 2
10. Ratha NK, Chen S, Jain AK (1995) Adaptive flow orientation-based feature extraction in fingerprint images. Pattern Recognit 28(11):1657–1672
11. Mehtre BM (1993) Fingerprint image analysis for automatic identification. Mach Vision Appl 6:124–139
12. Farina A, Kovács-Vajna ZM, Leone A (1999) Fingerprint minutiae extraction from skeletonized binary images. Pattern Recognit 32(5):877–889
13. Sagar VK, Ngo DBL, Foo KCK (1995) Fuzzy feature selection for fingerprint identification. In: IEEE 29th annual 1995 international Carnahan conference on security technology, Sanderstead, 18–20 Oct 1995
14. Deutsch ES (1972) Thinning algorithm on rectangular, hexagonal and triangular arrays. Commun ACM 15(9):827–837
15. Sagar VK, Berstecher RG (1994) Fuzzy control for feature extraction from fingerprint images. In: Second European congress on intelligent techniques and soft computing (EUFIT94), Aachen, Germany, 20–23 Sept 1994
16. O'Gorman L, Nickerson JV (1989) An approach to fingerprint filter design. Pattern Recognit 22(1):29–38
17. Xiao Q, Raafat H (1991) Fingerprint image postprocessing: a combined statistical and structural approach. Pattern Recognit 24(10):985–992
18. Zhao F, Tang X (2007) Preprocessing and postprocessing for skeleton-based fingerprint minutiae extraction. Pattern Recognit 40:1270–1281
19. Kumar A et al (2013) Fuzzy binary decision tree for biometric based personal authentication. Neuro Comput 99:87–97
20. Hasan H, Abdul-Kareem S (2013) Fingerprint image enhancement and recognition algorithms: a survey. Neural Comput Appl 23:1605–1610

21. Kamei T, Mizoguchi M (1995) Image filter design for fingerprint enhancement. In: Proceedings of the international symposium on computer vision, pp 109–114
22. Hsieh CT, Lai E, Wang YC (2003) An effective algorithm for fingerprint image enhancement based on wavelet transform. Pattern Recognit 36(2):303–312
23. Bansal R, Sehagal P, Bedi P (2010) Effective morphological extraction of true fingerprint minutiae based on the hit or miss transform. Int J Biometrics Bioinf, 4(2):71–85
24. Espinosa V (2002) Mathematical morphological approaches for fingerprint thinning. IEEE
25. Rutovitz D (1966) Pattern recognition. J Roy Stat Soc 129:504–530

Hybrid Model for Analysis of Abnormalities in Diabetic Cardiomyopathy

Fahimuddin Shaik, Anil Kumar Sharma and Syed Musthak Ahmed

Abstract Nowadays Image processing methods have become indispensable in solving various medical Imaging problems. Cardiac problems are main cause for 80 % of deaths in Diabetic patients. The proposed work is mainly dealt with processing of medical images related to Diabetic Cardiomyopathy. The motto of this paper is on observing enhancement and segmentation of the cross sectional view of a blood capillary of a right coronary artery image of a diabetic patient. In this work the Results and Analysis are derived from applying Hybrid Morphological Reconstruction Technique as Pre-Processing with Watershed Segmentation Method as Post-Processing.

Keywords Image · Segmentation · Artery · Morphological · Watershed

1 Introduction

Since Human Life is worthier than all things, much effort has been carried out today to diagnose a disease and its disorders. Out of many disorders Diabetic mellitus (DM) is a metabolic disorder that is characterized by inability of the pancreas to

F. Shaik (✉)
Electronics and Communication Engineering, SunRise University, Alwar, Rajasthan, India
e-mail: fahimuddin.shaik.in@ieee.org

A.K. Sharma
Institute of Engineering and Technology, Alwar, Rajasthan, India
e-mail: aks_826@yahoo.co.in

S.M. Ahmed
Department of Electronics and Communication Engineering, SREC, Warangal, Telangana, India
e-mail: syedmusthak_gce@rediffmail.com

R. Bhramaramba and A.C. Sekhar (eds.), *Application of Computational Intelligence to Biology*, Springer Briefs in Forensic and Medical Bioinformatics,
DOI 10.1007/978-981-10-0391-2_5

control blood glucose concentration. This predicament results may make out blood glucose levels out of range [1]. Cardiac problem is responsible for 80 % of deaths among diabetic patients much of which has been ascribed to CAD (coronary artery disease). Nonetheless, there is an increase in recognition that diabetic patients have a medical condition from an additional cardiac insult termed Diabetic Cardiomyopathy [2]. Diabetic Cardiomyopathy (also termed as Atherosclerosis) is a condition in which an artery wall thickens as the result of an accumulation of fatty materials such as cholesterol. The foremost pathology in diabetic patients is that basement membrane of a blood vessel gradually thickens making blood vessel narrower [3]. Epidemiological and medical trial data have established the larger incidence and prevalence of heart attacks in diabetes even without a noise at all [4].

On a technical note Image Processing is utilised to extract important features from the images, through which better perception of the scene can be obtained for human viewers [5]. The biological vision system is one of the most important means of exploration of the world to humans, making complex task easier for betterment of understanding [6].

There are numerous algorithms that can be utilised for different applications but enhancement and segmentation are considered as most sort out methods for improving the details in an image. It is not possible to judge that any one method is best in Image processing applications but one can use trial and error method as a practical approach for obtaining the perfect results. Image Enhancement is a fundamental task in digital image processing and analysis, aiming to improve the appearance of image in terms of human brightness perception [7]. Whereas the Segmentation is mainly useful in classification of objects and labelling of the features extracted from image for easy analysis. One should look into that processing of images is done without blemishing the integrity of original image.

2 Hybrid Morphological Reconstruction

Due to the imperfection and variations, the appearance of microscopic images is generally not homogeneous. In order to reduce the influence from undesirable variations within, the Hybrid Morphological Reconstruction (HMR) [8] is used to enhance the image.

3 Watershed Segmentation

Watershed segmentation falls under Morphological image processing methods and is a distinguished image segmentation technique because of its significance related to mathematical morphology. Morphological operators have been applied

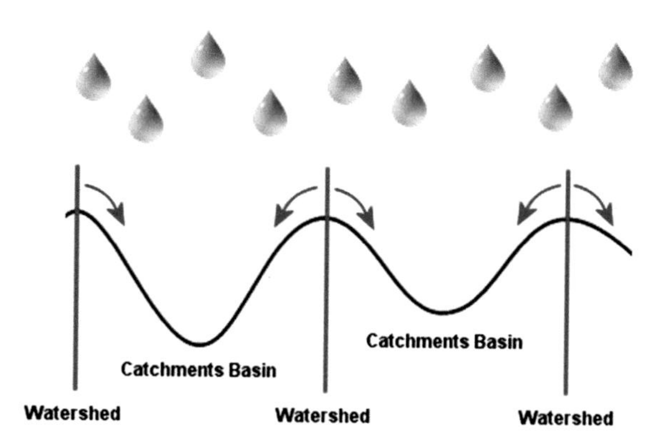

Fig. 1 Watershed representation

for vasculature segmentation [9] because the fundamental morphology of the vasculature is known a priori to be comprised of linked linear segments and because of speed and noise resistance. The concept of watersheds is based on visualizing an image in 3 dimensions given by two spatial co-ordinates versus intensity. Here one can consider only three points for clear explanation of the topic and they are (a) points belonging to regional minimum (b) points at which a drop of water, if placed at the location of any of those points, would fall with indeed to a single minimum; and (c) points at which water would be equally likely to fall to more than one such minimum. The points fulfilling condition (c) form crest lines on the topographic surface and are termed *divide lines* or *watershed lines* [10] (Fig. 1).

The standard objective of this method is based on the concept to find watershed lines. The basic idea is simple, suppose that a hole is punched in each regional minimum and that the entire geography is flooded from below by allowing water rise through the holes at uniform rate [11]. The entire process is described by a concept that a dam like thing is constructed to avoid merging and flooding may take place when water reaches the top level of dam. Consequently, watershed algorithm extracts the boundaries. In [12] watershed algorithm was used for segmentation of splats, a collection of pixels with similar color and spatial location.

4 Existing Work

In literature [3, 8] the above said algorithms have been used as pre-processing or post-processing methods with other algorithms. But in this work these two algorithms are combined to form a hybrid model to attain effective results for easy classification and analysis of the inner lying cause of the anomalies present in the images related to Diabetes (Fig. 2).

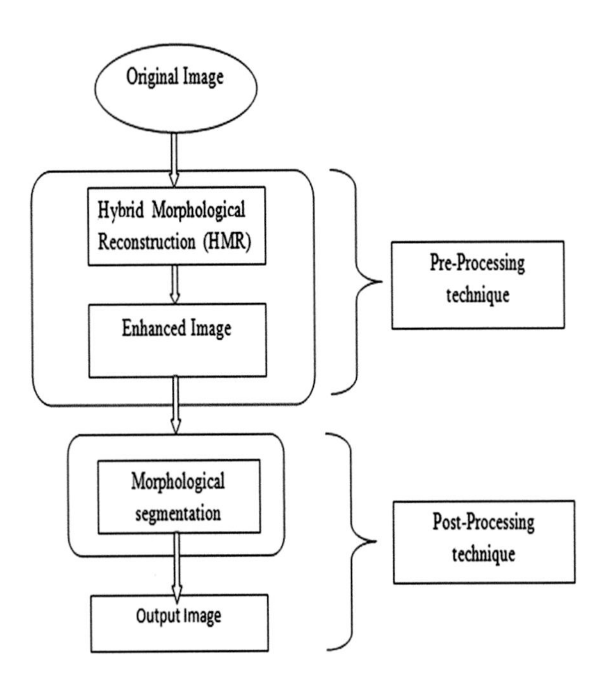

Fig. 2 Block diagram of implemented model

5 Proposed Work

The implemented method is used to enhance and segment the medical image. Here Hybrid Morphological Reconstruction (HMR) technique is used to enhance the medical images, and Morphological segmentation is the enhanced medical image.

5.1 Model Description

The images that are obtained from the Nikon Database is taken as the input image. Here we have taken the image of cross section of the Right Coronary Artery (RCA) which carries pure blood. Generally the image is RGB image. So the RGB image is processed using the MATLAB software and the image undergoes several algorithms to get a better output. Initially the RGB image is converted into grey scale to avoid complex calculations. Next step is to perform the Gradient Magnitude segmentation function. After the above two steps are finished then the main step, watershed transform segmentation is performed. Watershed transform is the region base segmentation method. In this step it fills the gaps present in the images and finally the analyzing the result.

6 Result and Analysis

6.1 Normal Image

Figure 3 represents the cross sectional view of a blood capillary of a right coronary artery. Arteries have three layers inter most layer tunica intima, middle layer tunica media and outermost layer tunica adventitia. This is the artery of the healthy person.

Figure 4 represents the gray scale image of the original image. Here the pixel values varies from 0 to 255. The operations like Top-hat and Bottom-hat cannot be performed directly on the color images. So in order to perform these pre-processing methods, the RGB image is converted into the gray scale image.

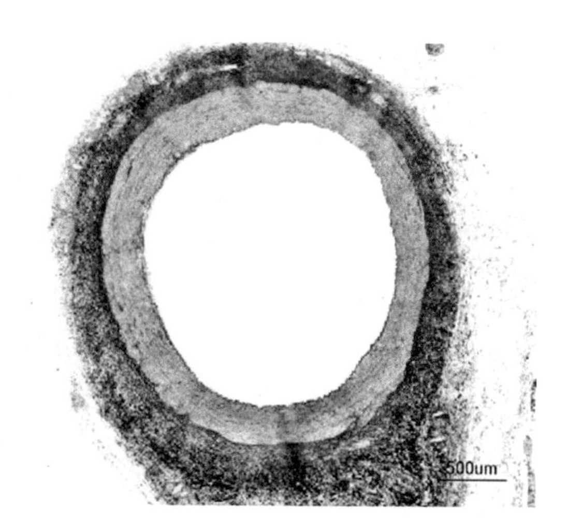

Fig. 3 Input image

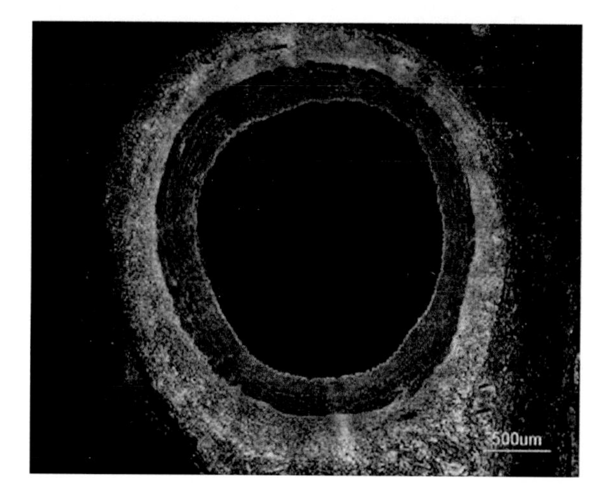

Fig. 4 Gray scale image

Sobel filtering is used for the clear detection of edges of the accumulated area. This along with Water shed-Gradient Magnitude gives better results for human perception. Figure 5 clearly gives the forecasting of how lumen will be occluded. Here Sobel edge detection method and a few arithmetic operators are used for better results.

Figure 6 shows the watershed transform on gradient magnitude image. Sorting out touching objects in an image is one of the most tricky image processing operations. In this figure the three layers are legibly seen with clear boundaries separated.

Figure 7 shows the opening by reconstruction image where Opening is erosion followed by dilation, here removal of pixels is done by erosion process and addition of pixels is done by dilation process. Opening removes small objects from

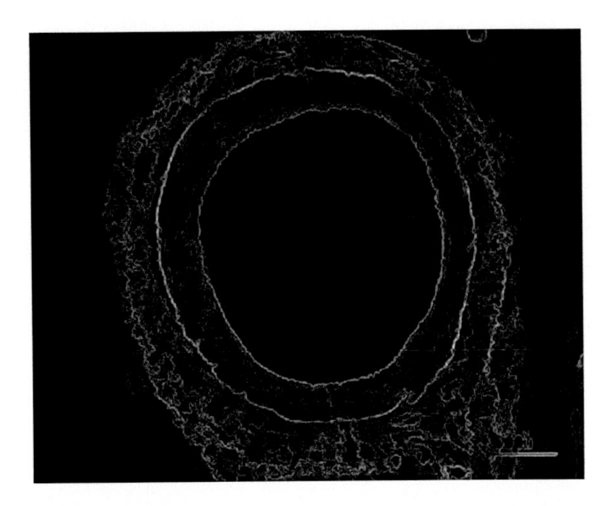

Fig. 5 Gradient magnitude image

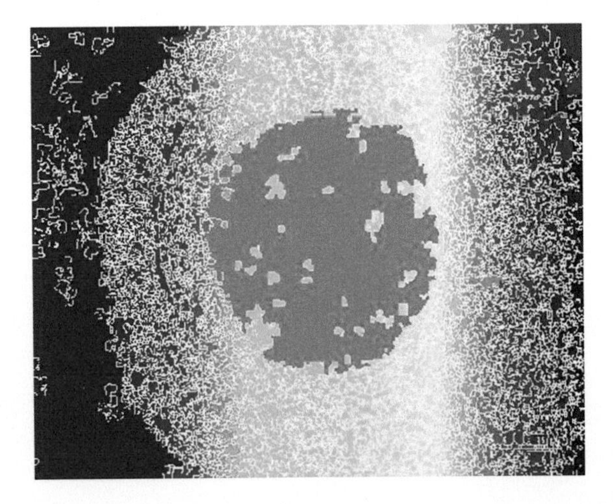

Fig. 6 Watershed transform of gradient magnitude

the fore ground of an image. Opening by reconstruction is erosion followed by a Morphological reconstruction, where Morphological reconstruction is repeated dilations of an image.

Figure 8 represents the reconstruction based on Opening and Closing image. The Opening with closing can remove the dark spots and stem marks. Opening-Closing reconstruction based operation is more effective than standard Opening and Closing at removing small blemishes without affecting the overall shapes of the objects.

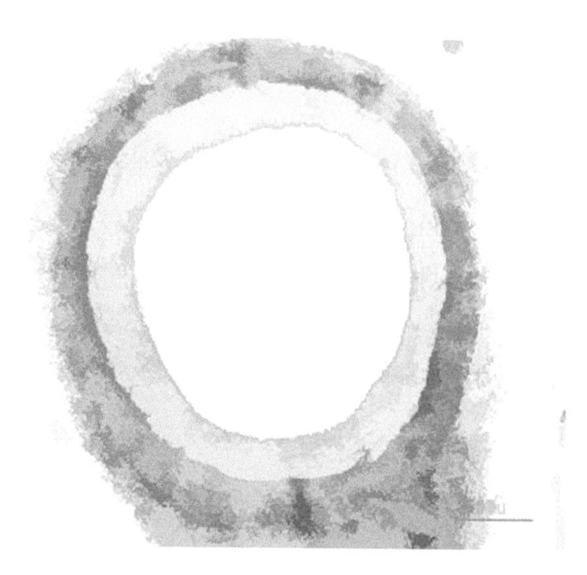

Fig. 7 Opening by reconstruction

Fig. 8 Opening-closing by reconstruction

Figure 9 shows the Regional maxima of a opening closing by reconstruction. Regional Maxima is a Morphological Operation which shows connected components of pixels with a constant intensity value and whose external boundary pixels all have a lower value. Regional Maxima is used to obtain good foreground markers.

Figure 10 shows the superimposed image on the original image from which we obtain the problem in the image. The analysis is done for healthy person artery and hence it gives the clear analysis for all layers present in the artery.

Fig. 9 Regional maxima of opening-closing by reconstruction

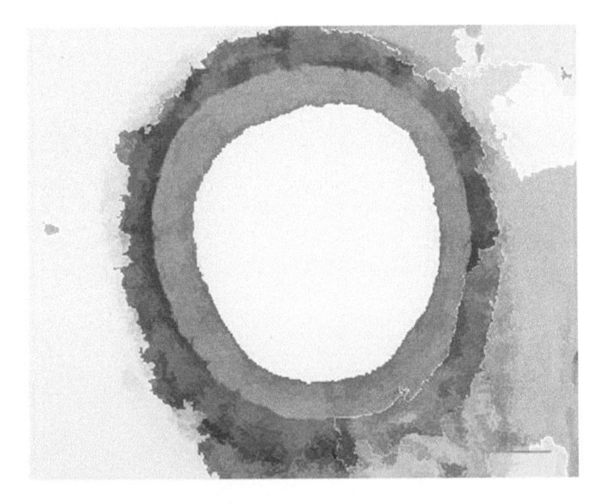

Fig. 10 Superimposition on original image

6.2 Medium Condition

Figure 11 represents the cross sectional view of a blood capillary of a right coronary artery under the medium condition. Atherosclerosis leads to narrowing of blood vessel which carries blood to the heart, thereby causing heart failure.

The operations like Top-hat and Bottom-hat cannot be performed directly on the color images. So in order to perform these pre-processing methods, the RGB image is converted into the gray scale image. Figure 12 represents the gray scale image of the original image. Here the pixel values vary from 0 to 255. Figure 13 shows the gradient magnitude of the image. Sobel filtering is used for the clear

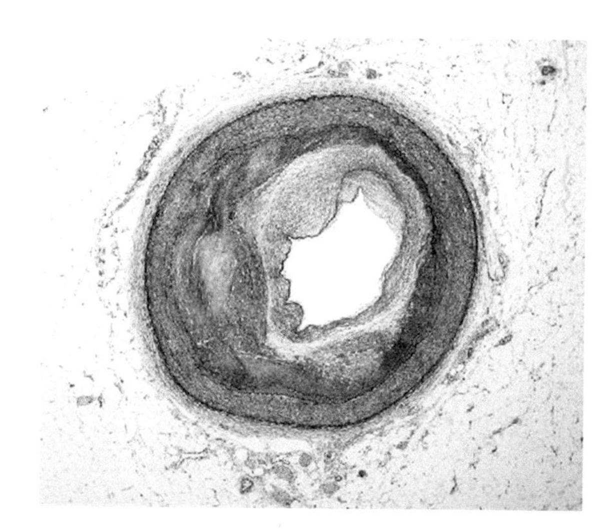

Fig. 11 Input image

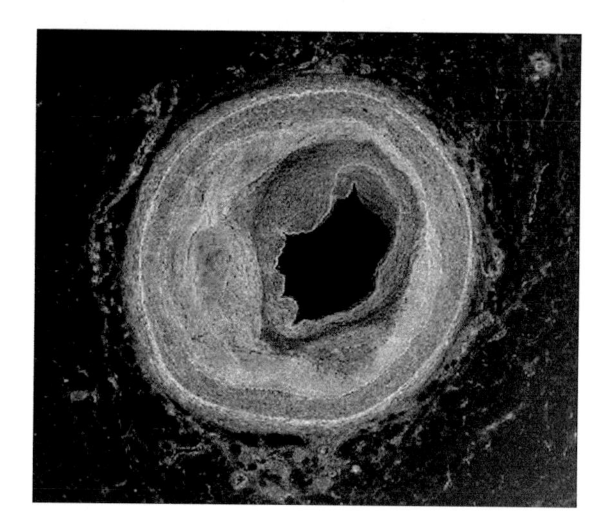

Fig. 12 Gray scale image

detection of edges of the accumulated area. This along with Watershed-Gradient Magnitude gives better results for human perception. The figure clearly gives the forecasting of how lumen will be occluded. Here Sobel edge detection method and a few arithmetic operators are used for better results.

Figure 14 shows the watershed transform of the image. From this figure one can easily state that the remaining lumen after occlusion with fatty materials is seen as a left out region in mid region of the image. Figure 15 shows the opening by reconstruction of the image. Where Opening is erosion followed by dilation, here removal of pixels is done by erosion process and addition of pixels is done by dilation process. Opening removes small objects from the fore ground of an image.

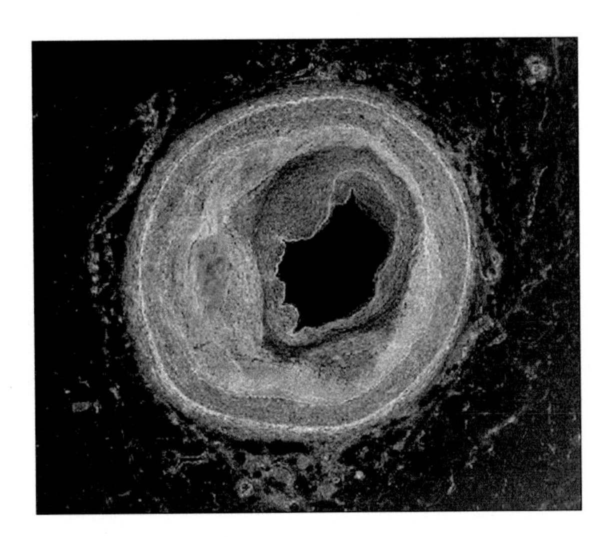

Fig. 13 Gradient magnitude

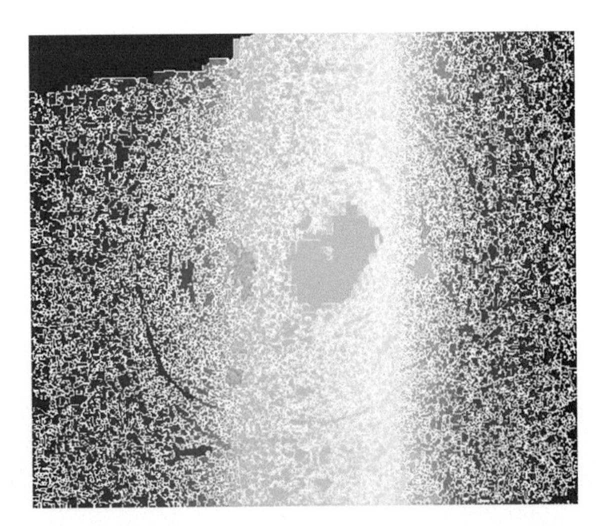

Fig. 14 Watershed transform

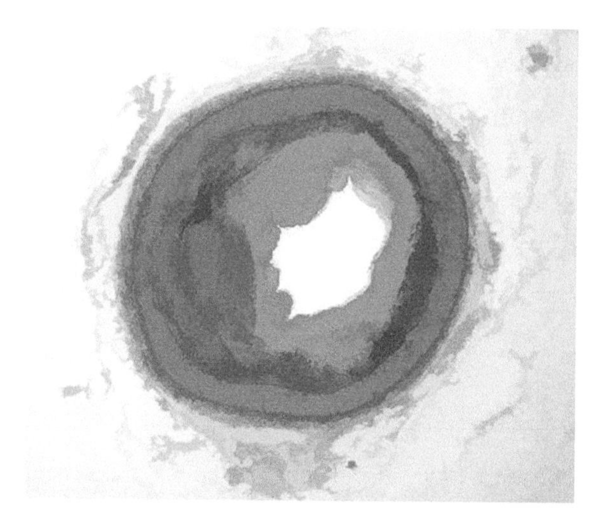

Fig. 15 Opening by reconstruction

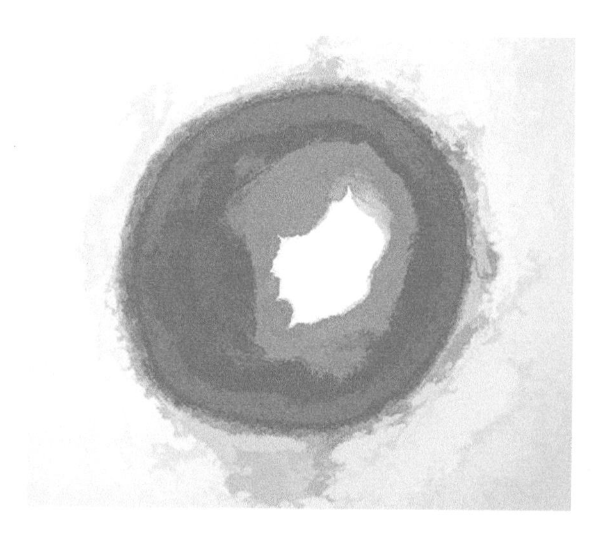

Fig. 16 Opening closing by reconstruction

Figure 16 shows the opening closing by reconstruction. The combinational operation of Opening-Closing reconstruction is more effectual than standard Opening and Closing at eliminating small blemishes without upsetting the overall shapes of the objects.

Figure 17 shows the regional maxima of opening closing by reconstruction. Regional Maxima is a Morphological Operation which shows connected components of pixels with a constant intensity value and whose external boundary pixels all have a lower value. Regional Maxima is used to obtain good foreground markers. By applying all these operations we obtained the left area of Lumen which is used for the flow of blood. By prolonging this process, we can educate the common man that if they neglect the medication what will be the health condition.

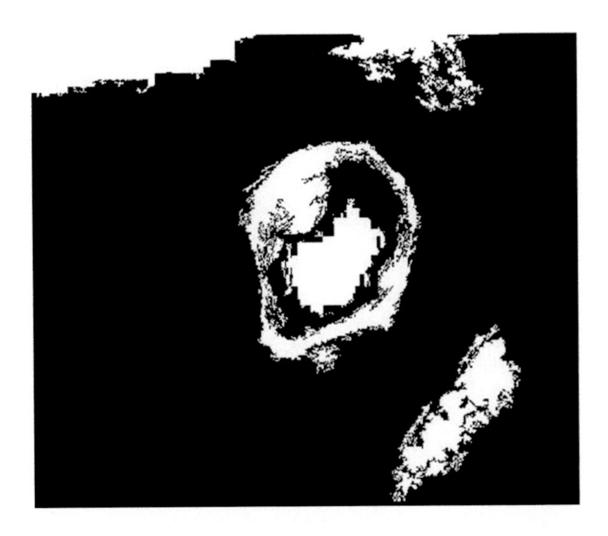

Fig. 17 Regional maxima of opening closing by reconstruction

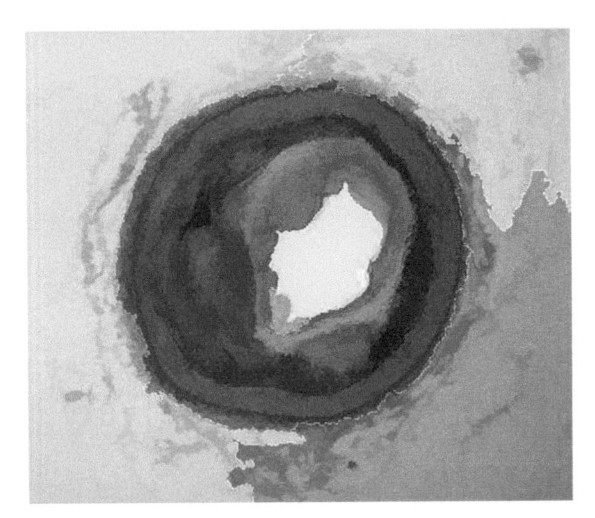

Fig. 18 Superimposed image on original

Figure 18 shows the superimposed image on the original image from which we obtain the problem in the image. It shows clearly shows how much part is left open for blood flow.

6.3 Severe Condition

Figure 19 represents the cross sectional view of a blood capillary of a right coronary artery. The problem in the figure is Atherosclerosis leading to narrowing of blood vessel which carries blood to the heart, thereby causing heart failure.

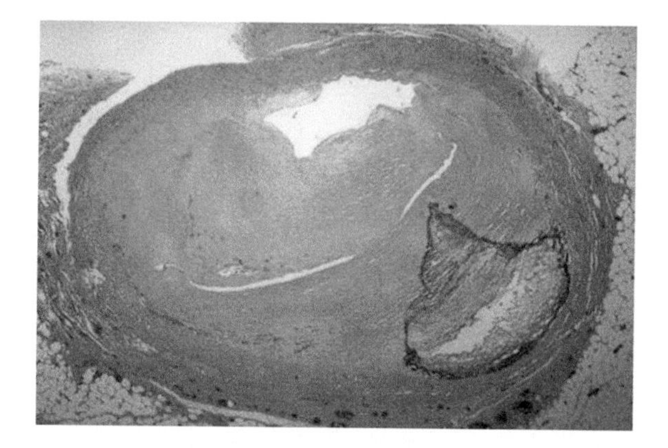

Fig. 19 Input image

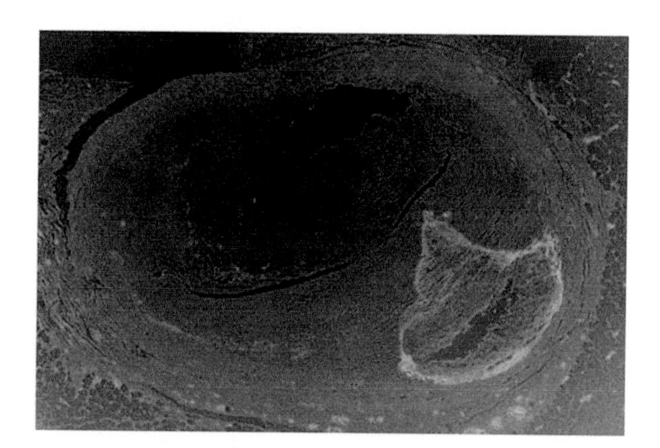

Fig. 20 Gray scale image

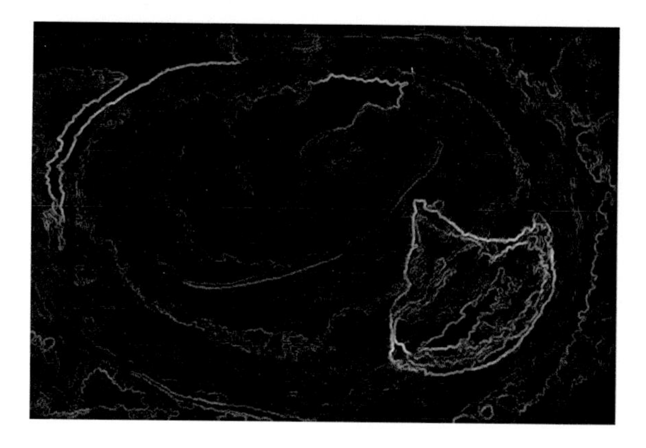

Fig. 21 Gradient magnitude image

In the above figure, the inner most layer i.e., Tunica Intima is severely thickened. Figure 20 represents the gray scale image of the original image. Here the pixel values vary from 0 to 255. Figure 21 shows the gradient magnitude of the image. Sobel filtering is used for the clear detection of edges of the accumulated

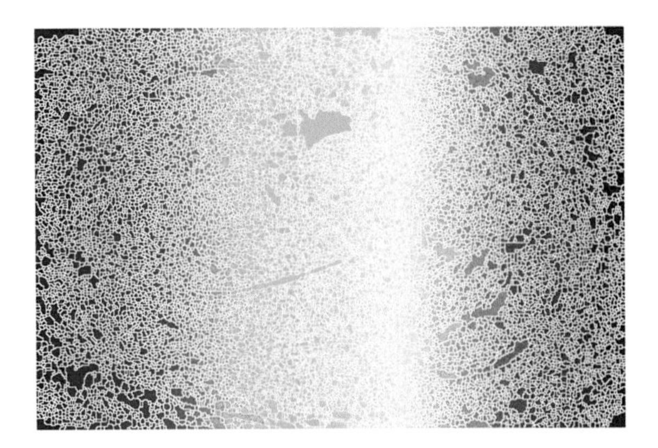

Fig. 22 Watershed transform of image

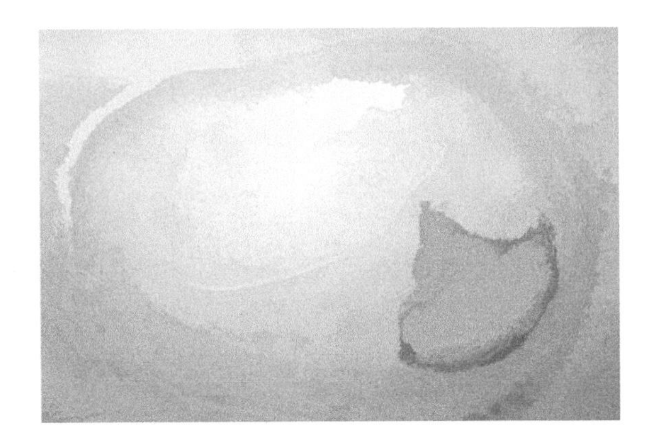

Fig. 23 Opening by reconstruction image

area. This along with Watershed-Gradient Magnitude gives better results for human perception. Here Sobel filtering and masks are utilized for better results. The figure clearly gives the forecasting of how lumen will be occluded.

Figure 22 shows the watershed transform of the image. Sorting out touching objects in an image is one of the more difficult image processing operations. As watershed is applied here it is evident from gradient image that affected part is legibly seen. Figure 23 shows the opening by reconstruction of the image. Where Opening is an erosion followed by dilation as discussed in earlier result of Fig. 15.

Figure 24 represents the reconstruction based on Opening and Closing image. Opening-Closing reconstruction based operation is more effective than standard Opening and Closing at get rid of small blemishes without upsetting the overall shapes of the objects. Figure 25 represents the regional maxima of the image. Regional Maxima is a Morphological Operation which shows connected components of pixels with a constant intensity value and whose external boundary pixels all have a lower value. Regional Maxima is used to obtain good foreground markers. By applying all these operations we obtained the left area of Lumen which is

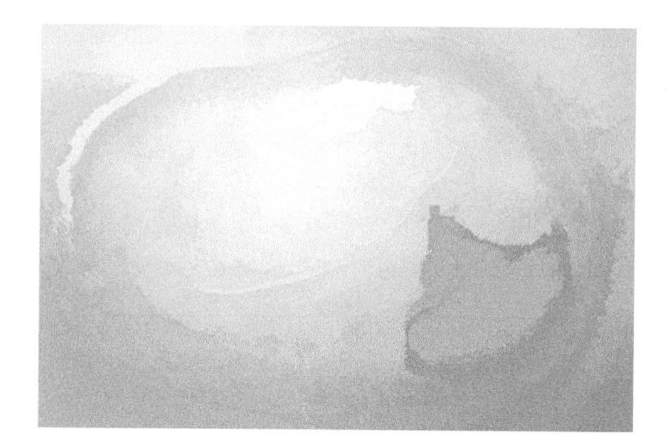

Fig. 24 Opening-closing by reconstruction

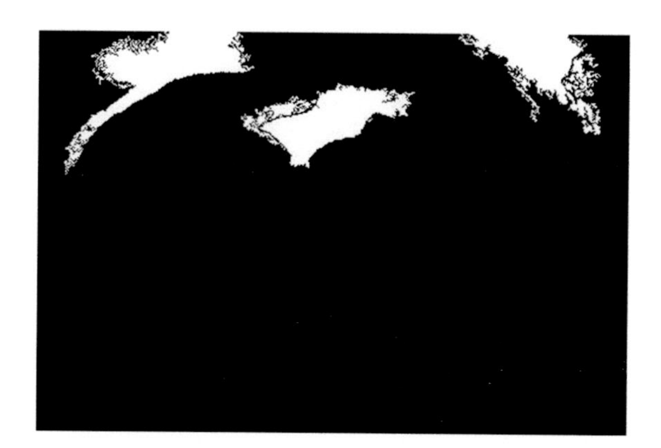

Fig. 25 Regional maxima output image

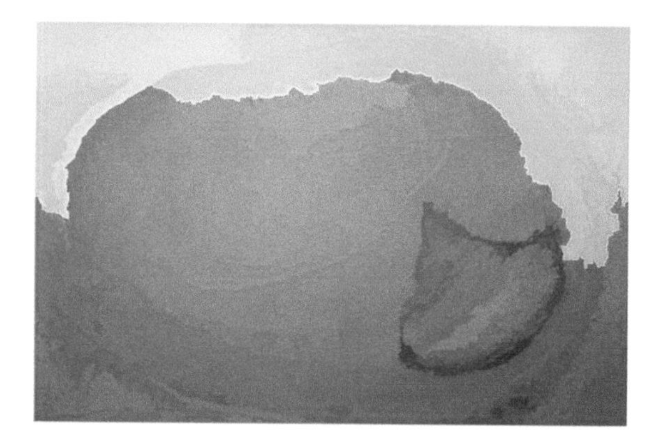

Fig. 26 Superimposed on original image

used for the flow of blood. By prolonging this process, we can educate the common man that if they neglect the medication what will be the health condition.

Figure 26 shows the superimposed on the original image from which we can obtain the problem in the image.

7 Statistical Analysis

7.1 Statistical Analysis Using MIPAV

The MIPAV open source tool facilitates quantitative analysis and visualization of medical images. MIPAV also lets researchers to perform statistical procedures on ROIs (Region-of-interest).

The performance of the implemented method was rigorously evaluated using quality metrics like Area, perimeter, Median, Standard Deviation of Intensity, Coefficient of skewness. By observing area, perimeter and Standard Deviation of Intensity, median, Coefficient of skewness the changes occurred in ROI of image after processing is easily revealed. Decrease in area indicates that ROI i.e., the degree of severity increases. Decrease in perimeter indicates that ROI i.e., the degree of severity increases. Increase in the standard deviation indicates that ROI i.e. has been detected with fine edges. The variation in skewness (decrease or increase) gives the asymmetry of a distribution. Increase in Median shows the Average changes of the pixels that occurred in the segmented image. The above all parameters are obtained by using the Medical Image Processing and Visualization (MIPAV) software are listed in Table 1. By observing the parameters like area and perimeter and then compare the input and out images in each condition i.e., normal, medium, and severe we clearly understand the problem of the images. The calculations are done only to the input and output images but not to each and every image that are obtained in the analysis. For example if we consider the area for three conditions, it will decreased and can be observed from the above table. This means that the gap of the valve is decreased and affects the flow of blood. Perimeter also decreases stage by stage.

Table 1 Quality assessment metrics for input and output images (ROI) under normal, medium and normal conditions

Parameters	Normal condition		Medium condition		Severe condition	
	Input image	Output image	Input image	Output image	Input image	Output image
Area	19,971	20,878	5972	6522	3650	3739
Perimeter	576.4831	544.9143	380.627	384.4449	344.3487	350.8232
Standard deviation in RGB	6.9521	5.2928	25.9016	18.8941	6.5016	6.6787
	11.3532	5.2928	24.8035	29.779	15.4811	2.8289
	7.5985	5.2928	20.3622	16.0068	8.8062	6.6281
Skewness in RGB	−5.8341	−4.9663	−4.8262	−3.996	−3.4868	−10.1449
	−5.8166	−4.9663	−4.9189	−4.1371	−2.6438	−6.5082
	−5.6576	−4.9663	−5.24	−3.9746	−1.7176	−5.5639
Median in RGB	196	138	254	253	248	249
	179	214	254	254	234	250
	172	214	254	178	205	174

By showing all these parameters we can educate the normal people about the severity of the problem in three stages i.e. Normal, medium and severe conditions.

8 Conclusion

The experimental results of Hybrid Morphological Reconstruction Technique as Pre-Processing and with Watershed Segmentation Method as Post-Processing are quite suitable for forecasting of narrowing of Lumen in CVD Patients. In future development other pre-processing algorithms combined with the implemented post processing method can give perspective results so that the Medical Professionals may make use of this algorithm for earlier detection of the abnormality and this framework in the form of a group wise images may be used to educate a common about seriousness of the problem.

Acknowledgement The authors are thankful to SunRise University-Alwar, Rajasthan and Annamacharya Institute of Technology and Sciences, Rajampet, A.P. for providing research facilities. And also thankful to Dr. B. Jayabhaskar Rao, Diabetalogist, Diabetic Care Center Nandalur, A.P. for providing the detailed explanation of Diabetes and its abnormalities.

References

1. Sharifi A, Vosolipour A, Aliyari Sh M, Teshnehlab M (2008) Hierarchical Takagi-Sugeno type fuzzy system for diabetes mellitus forecasting. In: Proceedings of 7th international conference on machine learning and cybernetics, Kunming, vol 3, pp 1265–1270, 12–15 July 2008
2. Hayath SA, Patel B (2004) Diabetic cardiomyopathy: mechanisms, diagnosis, and treatment. Department of cardiology Northwick Hospital UK, Clinical Science
3. Kumar A, Shaik F (2015) Image processing in diabetic related causes. In: Springer Briefs in applied sciences and technology-forensics and medical bio-informatics. Springer, May 2015. ISBN: 978-981-287-623-2. http://www.springer.com/in/book/9789812876232#aboutBook
4. Asghar O, AL Sunni A, Withers S (2009) Diabetic cardiomyopathy. The Manchester heart centre, UK, Clinical Science
5. Gonzalez RC, Woods RE (2002) Digital image processing, 1st edn. Addison-Wesley, An imprint of Pearson Education
6. Peres FA, Oliveira FR, Neves LA, Godoy MF (2010) Automatic segmentation of digital images applied in cardiac medical images. In: IEEE-PACHE, conference, workshops, and exhibits cooperation, Lima, PERU, 15–19 Mar 2010
7. Intajag S, Tipsuwanporn V, Chatthai R Chatree (2009) Retinal image enhancement in multimode histogram. In: 2009 World congress on computer science and information engineering, vol 4, pp 745–749, Mar 2009
8. Lu C, Mahmood M, Jha N, Mandal M (2012) A robust automatic nuclei segmentation technique for quantitative histopathological image analysis. Anal Quant Cytol Histopathol, 296–308
9. Zana F, Klein J-C (1999) A multimodal registration algorithm of eye fundus images using vessels detection and hough transform. Med Imaging IEEE Trans 18(5):419–428

10. Shaik F, Giriprasad MN, Swathi C, Soma Sekhar A (2010) Detection of cardiac complications in diabetic patients using Clahe method. In: Proceedings of international conference on aerospace electronics, communications and instrumentation (Aseci-2010), India, 6–7 Jan 2010, pp 344–347
11. Ravindraiah R, Shaik F (2010) Detection of exudates in diabetic retinopathy images. In: National conference on "Future Challenges and Building Intelligent Techniques in Electrical and Electronics Engineering" (NCEEE' 10), Chennai, INDIA, July 2010, pp 363–368
12. Tang L, Niemeijer M, Abramoff MD (2011) Splat feature classification: detection of the presence of large retinal haemorrhages. In: 2011 IEEE international symposium on biomedical imaging: from nano to macro, IEEE, pp 681–684

Computational Screening of DrugBank DataBase for Novel Cell Cycle Inhibitors

Satyanarayana Kotha, Yesu Babu Adimulam and R. Kiran Kumar

Abstract The cell cycle is governed by a family of proteins i.e. cyclin dependent kinases (CDKs) and their inhibitors (CDKIs) through activating and inactivating events. Cdks are heteromeric serine/threonine kinases that control progression through the cell cycle in concert with their regulatory subunits, the cyclins. CDK activity was found to increase in different types of human tumors as well as progression and/or invasiveness of some cancers such as breast, leukemia, and melanoma. Computer aided drug design was implemented to find novel compounds as possible CDK inhibitors. Screening of 1400 drugs from DrugBank database was carried out using Molegro software against various CDKs derived from Protein Data Bank. From the result, the best three drugs which exhibited high binding affinity against all targets are reported. The analysis against eight proteins resulted in 14 drugs and the top three drugs obtained are Olmesartan, Verteporfin and Atorvastatin.

Keywords CDK · Cell cycle · Drugbank · Molegro · Molecular docking

S. Kotha
Department of Information Technology, Sir C. R. Reddy College of Engineering, Eluru 534007, India
e-mail: satyanarayana9.kotha@gmail.com

Y.B. Adimulam (✉)
Department of CSE, Sir C. R. Reddy College of Engineering, Eluru 534007, India
e-mail: yesubabuadimulam9@gmail.com

R. Kiran Kumar
Department of CS, Krishna University, Machilipatnam, India
e-mail: kirankreddi@gmail.com

R. Bhramaramba and A.C. Sekhar (eds.), *Application of Computational Intelligence to Biology*, Springer Briefs in Forensic and Medical Bioinformatics,
DOI 10.1007/978-981-10-0391-2_6

1 Introduction

The cell cycle is governed by a family of proteins i.e. cyclin dependent kinases (CDKs) and their inhibitors (CDKIs) through activating and inactivating events. CDK protein levels are constant throughout the cell cycle, whereas the cyclin protein levels are regulated by cell cycle process [1]. The cell-division cycle consists of four sequential phases—G1, S, G2, and M—and the transition between these phases is regulated by the coordinated activities of cyclin CDK complexes [2]. CDK2, CDK3, CDK4, and CDK6 activity allows cells to proceed through interphase, and CDK1 activity is required for cells to proceed through mitosis [3]. CDK activity was found to increase in different types of human tumors as well as progression and/or invasiveness of some cancers such as breast, leukemia, and melanoma [4].

In complex with cyclin A, CDK2 promotes progression through S phase, and the active CDK2 complex persists in the nucleus through G2. CDK2 is positively regulated by phosphorylation of a conserved threonine residue that lies within the activation loop. CDK2 activity was found to increase in different types of human tumors, and correlations were found between the expression of CDK2 and its cyclin partners [5, 6].

The activity of CDK4 is restricted to the G1-S phase, which is controlled by the regulatory D-type cyclins and CDK inhibitor p16. This kinase was shown to be responsible for the phosphorylation of retinoblastoma gene product (Rb). Melanomas with wild-type BRAF or NRAS frequently have increased gene copy number for CDK4 and cyclin D1, which are downstream components of the RAS/BRAF pathway [7, 8].

Cdk5 is mainly active in post-mitotic neurons. The activity of Cdk5 is regulated by its binding with a neuron-specific regulatory subunit, either p35 or its isoform p39; its activity is therefore correlated with the expression of p35 and p39. CDK5 is active in prostate cancer cell lines and in almost all human metastatic prostate cancers, and inhibition of CDK5 activity resulted in reduction of spontaneous metastases by 79 % [9].

Molecular docking is a key tool in structural molecular biology and computer-assisted drug design. The goal of ligand-protein docking is to predict the predominant binding mode(s) of a ligand with a protein of known three-dimensional structure. Successful docking methods search high-dimensional spaces effectively and use a scoring function that correctly ranks candidate dockings [4]. In this paper, an attempt has been made to dock DrugBank drugs against cell cycle proteins for possible inhibitory activity using Molegro docking software.

2 Materials and Methods

2.1 Selection of Targets

Proteins or enzymes that are involved in cell cycle, apoptosis and cancer pathways are selected from Protein Data Bank (www.rcsb.org/pdb). The RCSB PDB provides a variety of tools and resources for studying the structures of biological macromolecules and their relationship to sequence, function, and disease. The structures of the selected proteins (2W9F, 1UNH, 1UV5, 3G0E and others) (Table 1) were downloaded from the PDB [10].

2.2 *DrugBank Database*

The Drug Bank database (www.drugbank.ca) is a unique bioinformatics and cheminformatics resource that combines detailed drug (i.e. chemical, pharmacological and pharmaceutical) data with comprehensive drug target (i.e. sequence, structure, and pathway) information (Sakharkar et al. [11]. Of the 4800 drugs available in the drug bank database 1400 drugs which are approved by FDA are selected for present study.

3 Molegro Virtual Docker

Molegro Virtual Docker is an integrated platform for predicting protein—ligand interactions (Thomsen and Christensen [12]. Molegro Virtual Docker handles all aspects of the docking process from preparation of the molecules to determination of the potential binding sites of the target protein, and prediction of the binding modes of the ligands. The Molegro Virtual Docker (MVD) has been shown to yield higher docking accuracy than other state-of-the-art docking products MVD: 87 %, Glide: 82 %, Surflex: 75 %, FlexX: 58 % (Thomsen and Christensen [12].

3.1 *Methodology*

The protein and ligand molecules present in the PDB or Mol2 formats were imported into the workspace of the Molegro Virtual Docker software. The molecules were prepared using default parameters of MVD. The cavities present in the protein are detected by Detect Cavities option and the large cavity was selected as the binding site for the ligand while performing docking.

4 Results and Discussion

The structure hits of proteins involved in apoptosis, Cell cycle regulator proteins other few proteins were selected based on specific criteria and downloaded from Protein Data Bank, these proteins are docked using Molegro Virtual Docker for

Table 1 List of all Protein targets

PDB ID	Expt method	Resolution (Å)	Ligands	Title
2W9F	X-ray diffraction	2.85	4-[(4-Imidazo[1,2-A]pyridin-3-ylpyrimidin-2-yl)amino]benzene sulfonamide	Crystal structure of human CDK4 in complex with a D-type cyclin
1GFW	X-ray diffraction	2.80	1-methyl-5-(2-phenoxymethyl-pyrrolidine-1- sulfonyl)-1 h-indole-2,3-dione	The 2.8 angstrom crystal structure of caspase-3 (apopain or cpp32) in complex with an isatinsulfonamide inhibitor
1UA2	X-ray diffraction	3.02	adenosine-5′-triphosphate	Crystal structure of human cdk7 & its protein recognition properties
1UNH	X-ray diffraction	2.35	(z)-1 h,1′h-[2,3]biindolylidene-3,2′-dione-3-oxime	Structural mechanism for the inhibition of cdk5-p25 by roscovitine, aloisine and indirubin
1UV5	X-ray diffraction	2.80	6-bromoindirubin-3′-oxime	Glycogen synthase kinase 3 beta complexed with 6-bromoindirubin-3′-oxime
2AZ5	X-ray diffraction	2.10	6,7-dimethyl-3-[(methyl{2-[methyl({1-[3-(trifluoromethyl)phenyl]- 1 h-indol-3-yl}methyl)amino]ethyl}amino)methyl]- 4 h-chromen-4-one	Crystal structure of tnf-alpha with a small molecule inhibitor
2UZO	X-ray diffraction	2.30	4-{5-[(z)-(2,4-dioxo-1,3-thiazolidin-5-ylidene)methyl]furan- 2-yl}benzenesulfonamide	Crystal structure of human cdk2 complexed with a thiazolidinone inhibitor
3BLR	X-ray diffraction	2.80	2-(2-chloro-phenyl)-5,7-dihydroxy-8-(3-hydroxy- 1-methyl-piperidin-4-yl)-4 h-benzopyran-4- one	Crystal structure of human cdk9/cyclint1 in complex with flavopiridol
3GOE	X-ray diffraction	1.60	N-[2-(diethylamino)ethyl]-5-[(Z)-(5-fluoro- 2-oxo-1,2-dihydro-3H-indol-3-ylidene)methyl]- 2,4-dimethyl-1H-pyrrole-3-carboxamide	Kit kinase domain in complex with sunitinib

Table 2 Mol dock scores of cell cycle regulator proteins

S. No	PDB ID	Mol dock score (kcal/mol)			Average mol dock score (kcal/mol)	Average RMSD value (kcal/mol)
		1 Run	2 Run	3 Run		
1	2W9F	145.5	147.78	146.3	146.51	0.67758
2	2UZO	126.54	123.25	125.63	125.15	1.06333
3	1UNH	113.91	111.66	114	133.19	0.11708
4	3BLR	114.24	112.32	114.63	114.06	0.17279
5	1GFW	35.672	36.853	34.662	35.729	0.278322
6	1UV5	117.86	115.63	114.32	115.93	0.170907
7	2AZ5	107.61	105.33	105.19	105.19	0.35177
8	3G0E	132.71	133.2	135.69	133.86	0.186848

three times to obtain stability and average of three molecular dock scores along with average RMSD are given in the following Table 2. All docking parameters are set to default and the output is given in Table 3.

The docking protocol was validated before screening DrugBank database. All proteins with bound ligand was docked into the binding pockets of all CDKs to obtain the best docked pose and the RMSD (Root Mean Square Deviation) of all atoms between these two conformations were found to be <2.0 °A indicating that the default parameters of Molegro for docking simulation are reasonable in reproducing the X-ray crystal structure [13, 14].

The rationale behind selecting CDKs as potential cancer targets is based on published literature on these enzymes. Of all the cancer causing or participating proteins/enzymes such as CDKs, apoptotic proteins and others, it has been reported that the crucial phases of cell cycle can be arrested if any one of the CDKs are blocked in a way to reduce cell proliferation [15]. Hence, it has been postulated by many authors that targeting CDKs would provide a higher chance or rate of inhibiting cell cycle process [5]; Fischer et al. [9]. Docking of all 1400 drugs from DrugBank was carried out to evaluate best conformer based on the lowest docked energy (kcal/mol), in other words, it should possess highest affinity towards the binding site. Moreover, the virtual screening technique employed in this work recognized entirely diverse, yet specific drugs that bind in a comparable manner.

Post docking analysis and upon comparison of dock scores, it was evidenced that Verteporfin resulted in best dock scores against all enzymes in the study. This drug was found to show better binding affinity with all enzymes except 1GFW, which suggests the importance of structural features of Verteporfin in generating inhibitory properties against a set of targets. Moreover, the second and third best drugs were found to be Olmesartan and Atorvastatin, respectively. The 2-dimensional structures and docking images are given in Figs. 1, 2, 3 and 4.

Table 3 Docking results of MVD

Drug name	2W9F	2UZO	1UNH	1UA2	3BLR	1UV5	1GFW	2AZ5
Pentagastrin	158.351	149.198	145.323	153.772	156.789	174.485	157.735	142.698
Olmesartan	208.296	198.472	197.894	195.778	187.035	185.706	180.897	186.703
Teniposide	174.5	164.784	178.286	158.118	166.59	166.57	162.213	154.647
Verteporfin	200.02	223.223	185.378	214.107	213.67	207.713	174.257	170.052
Montelukast	175.412	167.435	155.934	174.441	163.284	163.861	159.996	146.489
Candoxatril	165.536	154.91	155.38	149.329	169.491	172.129	141.528	130.336
Pemetrexed	162.825	138.96	159.865	162.21	148.055	151.165	136.954	113.528
Losartan	168.644	163.567	161.244	154.625	168.133	156.338	164.782	123.683
Candesartan	170.916	166.307	160.613	166.715	167.591	157.993	171.369	144.118
Eprosartan	181.317	167.877	161.587	158.843	177.815	159.489	189.155	150.724
Tiagabine	171.856	138.006	130.721	147.229	136.154	130.713	143.947	115.993
Repoglinide	167.581	150.842	142.284	145.877	138.795	132.628	143.383	127.579
Telmisartan	175.608	187.981	166.982	172.328	189.228	175.854	175.086	146.46
Atorvastatin	183.618	189.203	183.401	183.588	191.542	193.999	189.093	175.936
Flavopiridol*	−100.38	−100.678	−98.922	−112.24	−134.456	−129.231	−138.497	−118.126
Dinaciclib*	−145.093	−149.187	−142.21	−138.133	−140.189	−120.342	−154.385	−148.88
Roscovitine*	−154.134	−115.478	−125.036	−146.154	−162.223	−132.242	−152.115	−138.257

All the values in table represent binding affinity/energy between drug versus protein in kcal/mol
*Standarddrugs

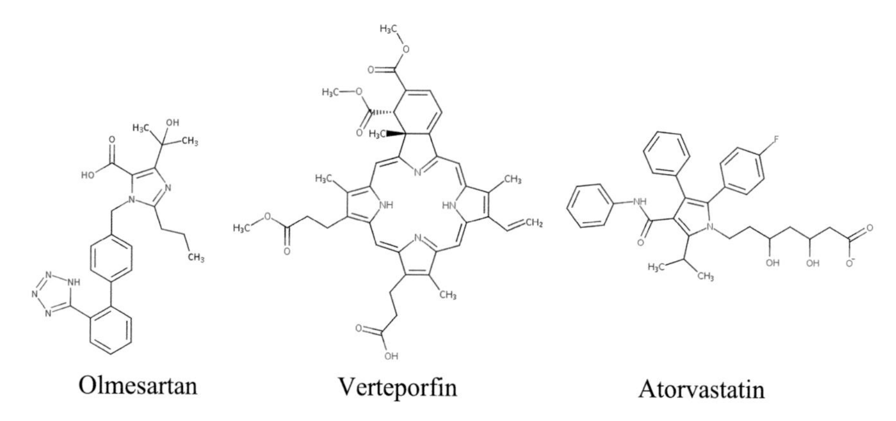

Fig. 1 2-dimensional structures of *top* three drugs

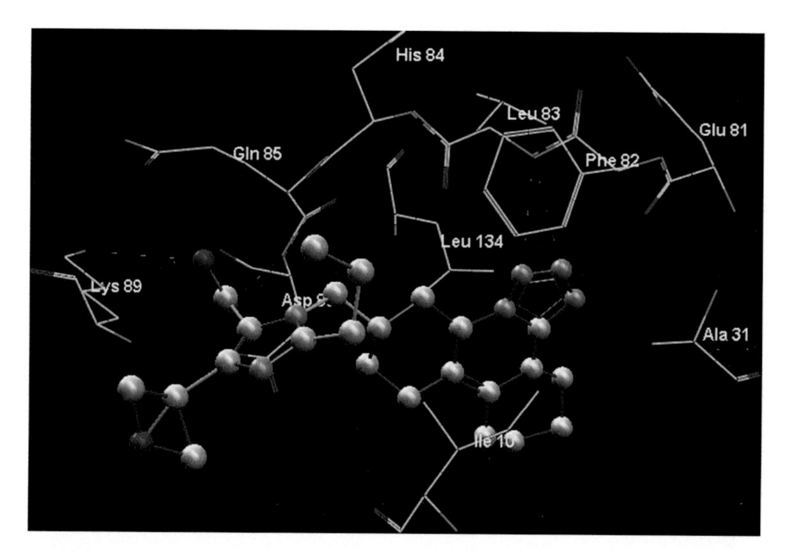

Fig. 2 Docked pose of Olmesartan with active site residues of 2UZO. H-bond interactions are shown in *dotted lines*

Finally, the best top three drugs which reported to exhibit high binding affinity are:

- 2W9F: Olmesartan, Verteporfin, Atorvastatin
- 2UZO: Olmesartan, Verteporfin, Atorvastatin
- 1UNH: Olmesartan, Verteporfin, Atorvastatin
- 1UA2: Olmesartan, Verteporfin, Atorvastatin
- 3BLR: Telmisartan, Verteporfin, Atorvastatin.

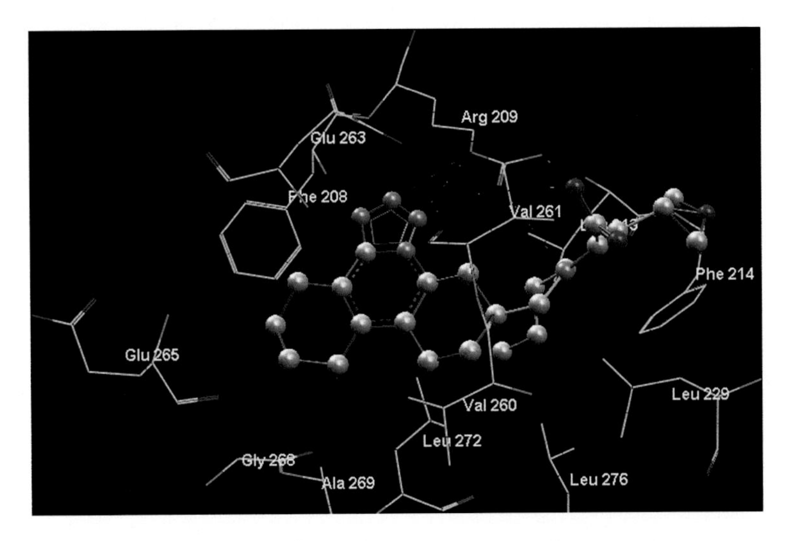

Fig. 3 Docked pose of Olmesartan with active site residues of 2W9F. H-bond interactions are shown in *dotted lines*

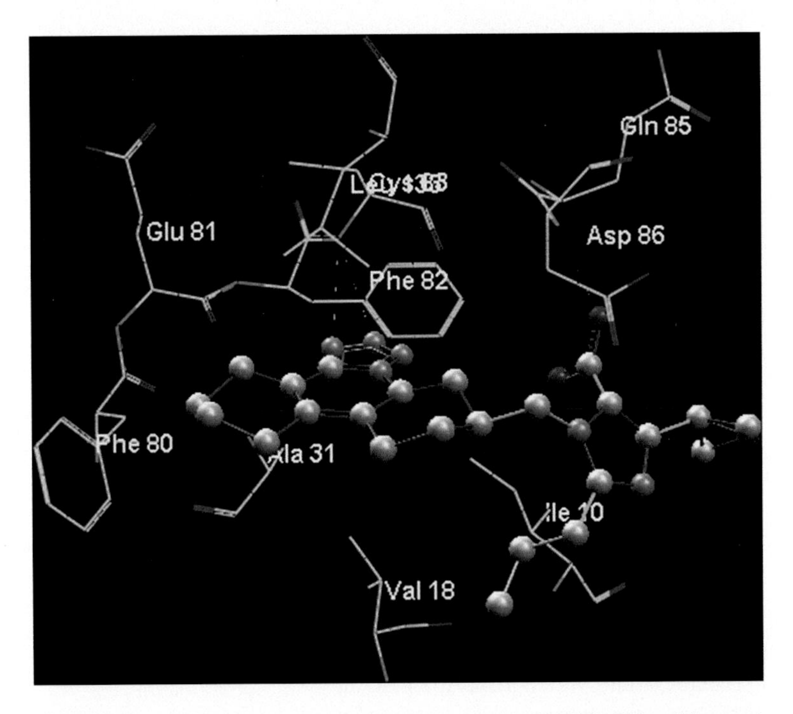

Fig. 4 Docked pose of Olmesartan with active site residues of 1UNH. H-bond interactions are shown in *dotted lines*

- 1UV5: Olmesartan, Verteporfin, Atorvastatin
- 1GFW: Olmesartan, Eprosartan, Atorvastatin
- 2AZ5: Olmesartan, Verteporfin, Atorvastatin

From the above data, it is evident that Olemsartan, Verteporfin and Atorvastatin exhibited moderate to high binding affinities against protein targets selected in the study.

Moreover, on the other hand, few known anti-cancer drugs such as Flavopiridol, Dinaciclib and Roscovitine are considered as positive control in the analysis and docking of these drugs against all proteins revealed dock scores much lower than few compounds obtained from analysis. Data given in Table 3 suggested that these standard drugs were able to generate weak binding energies when compared with the remaining compounds in the data set.

Therefore, further investigation of Olemsartan, Verteporfin and Atorvastatin as probable anti-cancer targets needs to be evaluated experimentally.

5 Conclusion

Screening studies of 1400 drugs obtained from drug bank database are docked against eight proteins of which one involved in apoptosis, four are involved in cell cycle progression and three are related to cancer, using Molegro Virtual Docker (MVD) software. After rigorous screening procedure, analysis resulted in 14 drugs and the top three drugs obtained are Olmesartan, Verteporfin and Atorvastatin. Hence this study employing computational molecular docking analysis reveals the importance of various drugs that are specific to a disease and may also be considered as possible anti-cancer agents.

References

1. Arris CE, Boyle FT, Calvert AH, Curtin NJ, Endicott JA, Garman EF, Gibson AE, Golding BT, Grant S, Griffin RJ, Jewsbury P, Johnson LN, Lawrie AM, Newell DR, Noble ME, Sausville EA, Schultz R, Yu W (2000) Identification of novel purine and pyrimidine cyclin-dependent kinase inhibitors with distinct molecular interactions and tumor cell growth inhibition profiles. J Med Chem 43:2797–2804
2. Choi YJ, Anders L (2014) Oncogene. Signaling through cyclin D-dependent kinases. 33. Issue 15:1890–1903
3. Brun V, Legraverend M, Grierson DS (2001) Cyclin-dependent kinase (CDK) inhibitors: development of a general strategy for the construction of 2,6,9-trisubstituted purine libraries. Part 1. Tetrahedron Lett 42:8161–8164
4. Gray N, Detivaud L, Doerig C, Meijer L (1999) ATP-site directed inhibitors of cyclin-dependent kinases. Curr Med Chem 6:859–875
5. Davies TG, Pratt DJ, Endicott JA, Johnson LN, Noble ME (2002) Structure-based design of cyclin-dependent kinase inhibitors. Pharmacol Ther 93:125–133

6. De Azevedo WF, Mueller-Dieckmann HJ, Schulze-Gahmen U, Worland PJ, Sausville E, Kim SH (1996) Structural basis for specificity and potency of a flavonoid inhibitor of human CDK2, a cell cycle kinase. Proc Natl Acad Sci USA 93:2735–2740
7. Endicott JA, Noble ME (2013) Structural characterization of the cyclin-dependent protein kinase family. Biochem Soc Trans 41:1008–1016
8. Endicott JA, Noble ME, Tucker JA (1999) Cyclin-dependent kinases: inhibition and substrate recognition. Curr Opin Struct Biol 9:738–744
9. Fischer PM, Endicott J, Meijer L (2003) Cyclin-dependent kinase inhibitors. Prog Cell Cycle Res 5:235–248
10. Helen MB (2007) The protein data bank: a historical perspective. Int Union Crystallogr A64:88–95
11. Sakharkar MK, Sakharkar KR, Pervaiz S (2007) Druggability of human disease genes. Int J Biochem Cell Biology 39:1156–1164
12. Thomsen R, Christensen MH (2006) MolDock: a new technique for high-accuracy molecular docking. J Med Chem 49:3315–3321
13. Adinarayana KPS, Ajay Babu P, Srinivas Kumar P (2012) In Silico lead identification by virtual screening and in vitro anti-cancer activities by MTT assay. Int J Computat Bioinfo Silico Model 1:55–57
14. Babu PA, Dondapati JS, Mangamoori LN, Srinivas K (2008) Identification of novel CDK2 inhibitors by QSAR and virtual screening procedures. QSAR Comb Sci 27:1362–1373
15. Sherr CJ (2000) The Pezcoller lecture: cancer cell cycles revisited. Cancer Res 60:3689–3695

Pathway Analysis of Highly Conserved Mitogen Activated Protein Kinases (MAPKs)

N. Deepak, Yesu babu Adimulam and R. Kiran Kumar

Abstract Mitogen activated protein kinases (MAPKs) are stimulated by a large variety of signals, including mitogens, growth factors, cytokines, T-cell antigens, pheromones, UV and ionizing radiations, osmotic stress, heat shock and oxidative stress. They participate in the generation of various cellular responses, including gene transcription, induction of cell death or maintenance of cell survival, malignant transformation, and regulation of cell-cycle progression. MAPKs are involved in the action of most nonnuclear oncogenes and responsible for cell response to growth factors. MAPK pathway has been shown to play a pivotal role in diverse dental diseases, including chronic pain, and periodontal diseases as well as in majority of various cancers. In this work, an attempt has been made to determine the participation of a particular MAPK in one specific pathway. Various computational analysis tools such as ClustalW, phylogenetic tree re-construction, PDB, Phosphosite etc. were utilized and based on evolutionary relationships, identification of phosphorylated sites and comparison of active site residues, the specificity of MAPK 1 and 3 in growth factor pathway, MAPK 8, 9, 10 in stress and MAPK 11, 12, 13 in inflammatory pathway are emphasized.

Keywords MAPK · Pathway · Phosphorylation · Active site residues

N. Deepak · Y.b. Adimulam (✉)
Department of CSE, Sir C R Reddy College of Engg,
Eluru, Andhra Pradesh, India
e-mail: yesubabuadimulam9@gmail.com

N. Deepak
e-mail: nedunurideepak@gmail.com

R. Kiran Kumar
Department of Computer Science, Krishna University,
Machilipatanam, Andhra Pradesh, India
e-mail: kirankreddi@gmail.com

R. Bhramaramba and A.C. Sekhar (eds.), *Application of Computational Intelligence to Biology*, Springer Briefs in Forensic and Medical Bioinformatics,
DOI 10.1007/978-981-10-0391-2_7

1 Introduction

The Mitogen-activated protein kinase (MAPK) family belongs to the eukaryotic protein kinase super family and identified by virtue of their activation in response to stimulation of cells [1–3]. The transmission of extracellular signals into intracellular medium is governed by a network of interacting proteins that regulate a large number of cellular processes. The signaling mechanism involves activation of several membrane receptors followed by a sequential stimulation of several MAPK signaling cascade [4, 5]. Another physiological response that appears to be regulated through the MAPK signaling pathway is cellular differentiation. Different members of the MAPK cascade have been implicated in processes such as monocytic differentiation, differentiation of PC12 cells, T cell maturation, and mast cell development. The activity of most MAPK's is stimulated by a large variety of signals, including mitogens, growth factors, cytokines, T cell antigens, pheromones, phorbol esters, UV and ionizing radiation, osmotic stress etc. [6, 7].

Mitogen-activated protein (Map) kinases are widely expressed serine-threonine kinases. Three major groups of Map kinases exist: the p38 Map kinase family, the extracellular signal-regulated kinase (Erk) family, and the c-Jun NH_2-terminal kinase (JNK) family. The members of the different Map kinase groups participate in the generation of various cellular responses, including gene transcription, induction of cell death or maintenance of cell survival, malignant transformation, and regulation of cell-cycle progression [8]. These signalling events ultimately regulate cellular responses such as proliferation, differentiation, secretion and apoptosis [8–10]. In general phosphorylation either activates or inactivates a given protein to perform a certain function. Protein kinases and phosphatases are responsible for determining the phosphorylation state of cellular proteins in the subcellular localization and activity of kinases and phosphatases together have consequences for normal cell function and maintenance of cellular homeostasis [11]. Extensive work by several groups has established that Map kinase pathways play critical roles in the pathogenesis of various hematologic malignancies, providing new molecular targets for future therapeutic approaches [12].

The rationale of this work is to perform multiple sequence analysis of MAPK involved in various pathways in order to ascertain the distribution of specific MAP Kinase pathways and to determine the specificity and participation in definite pathway by exploring the conserved and variant residues within active site regions of MAPKs.

2 Materials and Methods

MAP Kinase proteins involved in growth factor (Ras activation), stress (TGF activation) and inflammatory related pathways (Fig. 1) are taken from Biocarta. All protein sequences of MAPKs are extracted from ExPASy server. ClustalW multiple sequence alignment program was used to produce biologically meaningful multiple sequence alignment of divergent sequences (Figs. 2, 3, 4, Table 1).

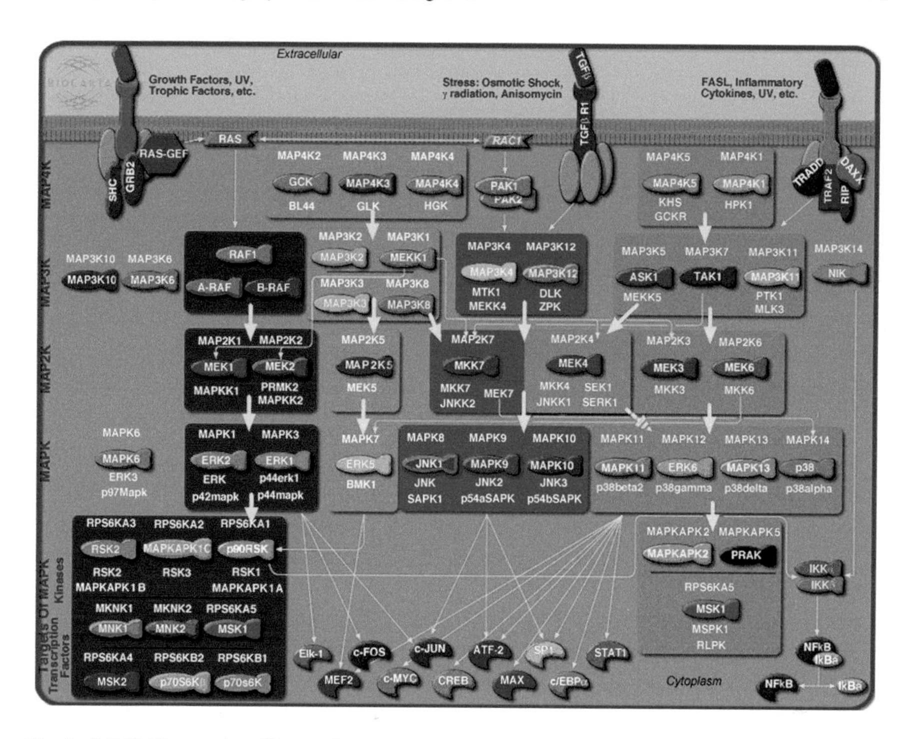

Fig. 1 MAP Kinase signalling pathway goes here

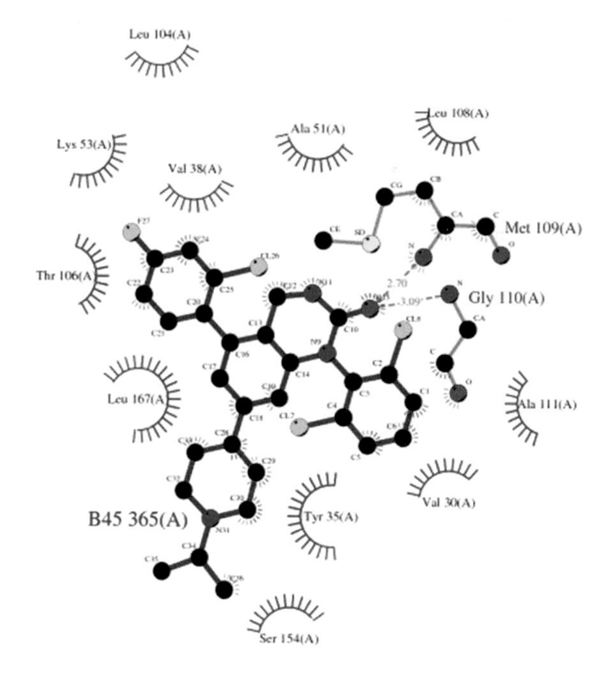

Fig. 2 Ligplot interactions between 3GC8 and bound ligand, B45. Hydrogen bonds are represented as *dotted lines* goes here

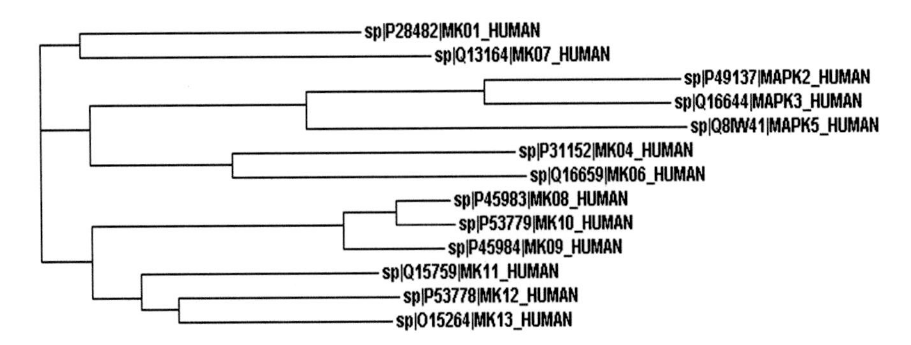

Fig. 3 Phylogenetic tree of MAPK's goes here

Phylogenetic relationships among MAPK proteins (Fig. 1) were carried out by using ClustalW alignment program. Proteins in fasta format are downloaded from swissprot protein sequence database and they are subjected to multiple alignment in clustalw with default parameters. The resulted alignment and trees are studied to evaluate evolutionary relationships and divergences among sequences. Phosphorylation sites data was given in Tables 2, 3, 4, 5, 6, 7, 8 and Figs. 5, 6, 7, 8, 9, 10, 11, 12.

Structural similarities within the active site region of all MAPKs are performed to analyze the conserved and variant residue patterns. Therefore, based on Ramachandran plot statistics, 3GC8 [13] is selected as reference protein and the structural features among other MAPKs are examined. Active site residues are listed based on Ligplot interactions (Fig. 2) and residue similarities at similar positions in other MAP kinases are identified. Further, multiple sequence analysis is performed to analyze variations within the active site regions.

3 Results and Discussion

The mitogen-activated protein kinase (MAP kinase) pathways consist of four major groups and numerous related proteins which constitute interrelated signal transduction cascades activated by stimuli such as growth factors, stress, cytokines and inflammation. The four major groups are the Erk (red), JNK or SAPK (blue), p38 (green) and the Big MAPK or ERK5 (light blue) cascades [14, 15]. The core unit of mitogen-activated protein kinase (MAPK) pathways is a three member protein kinase cascade. Within the three-kinase module, MAPKs are phosphorylated and activated by MAPK kinases (MKKs). The MKKs are phosphorylated and activated by serine/threonine kinases [16]. Activation of MAPKs in response to these stimuli controls gene expression, metabolism, cytoskeletal functions and other cellular regulatory events. The sequences of MAPK 1 to 13 which participate in the growth factor, stress, inflammatory pathway are analyzed and the phylogenetic tree was obtained (Fig. 3). A complete multiple sequence alignments of all MAPKs containing active domain regions was given in Fig. 4 and Table 1.

```
sp|P28482|MK01_HUMAN      -----------------------MAAAAAAG-----------AGP------EMVRGQVFD 20
sp|P27361|MK03_HUMAN      -----------------------MAAAAAQGGGGGEPRRTEGVGPGVPGEVEMVKGQPFD 37
sp|P53778|MK12_HUMAN      --------------------------------------MSSPPPARSGFYRQEVTKTAWE 22
sp|O15264|MK13_HUMAN      --------------------------------------MS--LIRKKGFYKQDVNKTAWE 20
sp|Q15759|MK11_HUMAN      --------------------------------------MS---GPRAGFYRQELNKTVWE 19
sp|Q16539|MK14_HUMAN      --------------------------------------MS---QERPTFYRQELNKTIWE 19
sp|P45983|MK08_HUMAN      --------------------------------------MSR-SKRDNNFYSVEIGDSTFT 21
sp|P53779|MK10_HUMAN      MSLHFLYYCSEPTLDVKIAFCQGFDKQVDVSYIAKHYNMSK-SKVDNQFYSVEVGDSTFT 59
sp|P45984|MK09_HUMAN      --------------------------------------MSD-SKCDSQFYSVQVADSTFT 21
sp|Q13164|MK07_HUMAN      ----------MAEPLKEEDGEDGSAEPPGPVKAEPAHTAASVAAKNLALLKARSFDVTFD 50
sp|P31152|MK04_HUMAN      ---------------------------------------------MAEKGDCIASVYGYD 15
sp|Q16659|MK06_HUMAN      ---------------------------------------------MAEKFESLMNIHGFD 15
                                                                                    :

sp|P28482|MK01_HUMAN      VGPRYTNLSYIGEGAYGMVCSAYDNVNKVRVAIKKIS-PFEHQTYCQRTLREIKILLRFR 79
sp|P27361|MK03_HUMAN      VGPRYTQLQYIGEGAYGMVSSAYDHVRKTRVAIKKIS-PFEHQTYCQRTLREIQILLRFR 96
sp|P53778|MK12_HUMAN      VRAVYRDLQPVGSGAYGAVCSAVDGRTGAKVAIKKLYRPFQSELFAKRAYRELRLLKHMR 82
sp|O15264|MK13_HUMAN      LPKTYVSPTHVGSGAYGSVCSAIDKRSGEKVAIKKLSRPFQSEIFAKRAYRELLLLKHMQ 80
sp|Q15759|MK11_HUMAN      VPQRLQGLRPVGSGAYGSVCSAYDARLRQKVAVKKLSRPFQSLIHARRTYRELRLLKHLK 79
sp|Q16539|MK14_HUMAN      VPERYQNLSPVGSGAYGSVCAAFDTKTGLRVAVKKLSRPFQSIIHAKRTYRELRLLKHMK 79
sp|P45983|MK08_HUMAN      VLKRYQNLKPIGSGAQGIVCAAYDAILERNVAIKKLSRPFQNQTHAKRAYRELVLMKCVN 81
sp|P53779|MK10_HUMAN      VLKRYQNLKPIGSGAQGIVCAAYDAVLDRNVAIKKLSRPFQNQTHAKRAYRELVLMKCVN 119
sp|P45984|MK09_HUMAN      VLKRYQQLKPIGSGAQGIVCAAFDTVLGINVAVKKLSRPFQNQTHAKRAYRELVLLKCVN 81
sp|Q13164|MK07_HUMAN      VGDEYEIIETIGNGAYGVVSSARRRLTGQQVAIKKIPNAFDVVTNAKRTLRELKILKHFK 110
sp|P31152|MK04_HUMAN      LGGRFVDFQPLGFGVNGLVLSAVDSRACRKVAVKKIA--LSDARSMKHALREIKIIRRLD 73
sp|Q16659|MK06_HUMAN      LGSRYMDLKPLGCGGNGLVFSAVDNDCDKRVAIKKIV--LTDPQSVKHALREIKIIRRLD 73
                          :         :* *  * * :*       .**:**:   :      ::: **: ::  .

sp|P28482|MK01_HUMAN      HENIIGINDIIR-APT--------IEQMKDVYIVQDLMETDLYKLLKT-QHLSNDHICYF 129
sp|P27361|MK03_HUMAN      HENVIGIRDILR-AST--------LEAMRDVYIVQDLMETDLYKLLKS-QQLSNDHICYF 146
sp|P53778|MK12_HUMAN      HENVIGLLDVFTPDET--------LDDFTDFYLVMPFMGTDLGKLMKH-EKLGEDRIQFL 133
sp|O15264|MK13_HUMAN      HENVIGLLDVFTPASS--------LRNFYDFYLVMPFMQTDLQKIMG--MEFSEEKIQYL 130
sp|Q15759|MK11_HUMAN      HENVIGLLDVFTPATS--------IEDFSEVYLVTTLMGADLNNIVKC-QALSDEHVQFL 130
sp|Q16539|MK14_HUMAN      HENVIGLLDVFTPARS--------LEEFNDVYLVTHLMGADLNNIVKC-QKLTDDHVQFL 130
sp|P45983|MK08_HUMAN      HKNIIGLLNVFTPQKS--------LEEFQDVYIVMELMDANLCQVIQ--MELDHERMSYL 131
sp|P53779|MK10_HUMAN      HKNIISLLNVFTPQKT--------LEEFQDVYLVMELMDANLCQVIQ--MELDHERMSYL 169
sp|P45984|MK09_HUMAN      HKNIISLLNVFTPQKT--------LEEFQDVYLVMELMDANLCQVIH--MELDHERMSYL 131
sp|Q13164|MK07_HUMAN      HDNIIAIKDILRPTVP--------YGEFKSVYVVLDLMESDLHQIIHSSQPLTLEHVRYF 162
sp|P31152|MK04_HUMAN      HDNIVKVYEVLGPKG---TDLQGELFKFSVAYIVQEYMETDLARLLEQ-GTLAEEHAKLF 129
sp|Q16659|MK06_HUMAN      HDNIVKVFEILGPSGSQLTDDVGSLTELNSVYIVQEYMETDLANVLEQ-GPLLEEHARLF 132
                          *.*:: : :::                    :   *:*   * ::* .::     :  ::   :

sp|P28482|MK01_HUMAN      LYQILRGLKYIHSANVLHRDLKPSNLLLN-TTCDLKICDFGLAR-VADPDHDHTGFLTEY 187
sp|P27361|MK03_HUMAN      LYQILRGLKYIHSANVLHRDLKPSNLLIN-TTCDLKICDFGLAR-IADPEHDHTGFLTEY 204
sp|P53778|MK12_HUMAN      VYQMLKGLRYIHAAGIIHRDLKPGNLAVN-EDCELKILDFGLAR-----QADS--EMTGY 185

sp|O15264|MK13_HUMAN      VYQMLKGLKYIHSAGVVHRDLKPGNLAVN-EDCELKILDFGLAR-----HADA--EMTGY 182
sp|Q15759|MK11_HUMAN      VYQLLRGLKYIHSAGIIHRDLKPSNVAVN-EDCELRILDFGLAR-----QADE--EMTGY 182
sp|Q16539|MK14_HUMAN      IYQILRGLKYIHSADIIHRDLKPSNLAVN-EDCELKILDFGLAR-----HTDD--EMTGY 182
sp|P45983|MK08_HUMAN      LYQMLCGIKHLHSAGIIHRDLKPSNIVVK-SDCTLKILDFGLAR-----TAGTSFMMTPY 185
sp|P53779|MK10_HUMAN      LYQMLCGIKHLHSAGIIHRDLKPSNIVVK-SDCTLKILDFGLAR-----TAGTSFMMTPY 223
sp|P45984|MK09_HUMAN      LYQMLCGIKHLHSAGIIHRDLKPSNIVVK-SDCTLKILDFGLAR-----TACTNFMMTPY 185
sp|Q13164|MK07_HUMAN      LYQLLRGLKYMHSAQVIHRDLKPSNLLVN-ENCELKIGDFGMARGLCTSPAEHQYFMTEY 221
sp|P31152|MK04_HUMAN      MYQLLRGLKYIHSANVLHRDLKPANIFISTEDLVLKIGDFGLAR-IVDQHYSHKGYLSEG 188
sp|Q16659|MK06_HUMAN      MYQLLRGLKYIHSANVLHRDLKPANLFINTEDLVLKIGDFGLAR-IMDPHYSHKGHLSEG 191
                          :**:* *::::*:* ::******.*: :.     *:* ***:**              ::

sp|P28482|MK01_HUMAN      VATRWYRAPEIMLNSKGYTKSIDIWSVGCILAEMLSNRPIFPGKHYLDQLNHILGILGSP 247
sp|P27361|MK03_HUMAN      VATRWYRAPEIMLNSKGYTKSIDIWSVGCILAEMLSNRPIFPGKHYLDQLNHILGILGSP 264
sp|P53778|MK12_HUMAN      VVTRWYRAPEVILNWMRYTQTVDIWSVGCIMAEMITGKTLFKGSDHLDQLKEIMKVTGTP 245
sp|O15264|MK13_HUMAN      VVTRWYRAPEVILSWMHYNQTVDIWSVGCIMAEMLTGKTLFKGKDYLDQLTQILKVTGVP 242
sp|Q15759|MK11_HUMAN      VATRWYRAPEIMLNWMHYNQTVDIWSVGCIMAELLQGKALFPGSDYIDQLKRIMEVVGTP 242
sp|Q16539|MK14_HUMAN      VATRWYRAPEIMLNWMHYNQTVDIWSVGCIMAELLTGRTLFPGTDHIDQLKLILRLVGTP 242
sp|P45983|MK08_HUMAN      VVTRYYRAPEVILG-MGYKENVDLWSVGCIMGEMVCHKILFPGRDYIDQWNKVIEQLGTP 244
sp|P53779|MK10_HUMAN      VVTRYYRAPEVILG-MGYKENVDIWSVGCIMGEMVRHKILFPGRDYIDQWNKVIEQLGTP 282
sp|P45984|MK09_HUMAN      VVTRYYRAPEVILG-MGYKENVDIWSVGCIMGELVKGCVIFQGTDHIDQWNKVIEQLGTP 244
sp|Q13164|MK07_HUMAN      VATRWYRAPELMLSLHEYTQAIDLWSVGCIFGEMLARRQLFPGKNYVHQLQLIMMVLGTP 281
sp|P31152|MK04_HUMAN      LVTKWYRSPRLLLSPNNYTKAIDMWAAGCILAEMLTGRMLFAGAHELEQMQLILETIPVI 248
sp|Q16659|MK06_HUMAN      LVTKWYRSPRLLLSPNNYTKAIDMWAAGCIFAEMLTGKTLFAGAHELEQMQLILESIPVV 251
                          :.*::**:*.::*.   *.: :*:*:.***:.*::    :* *  . :.*   ::

sp|P28482|MK01_HUMAN      SQEDLNCIINLKARNYLLSLPHKNKVPWNRLFPN------------ADSKALDLLDKMLT 295
sp|P27361|MK03_HUMAN      SQEDLNCIINMKARNYLQSLPSKTKVAWAKLFPK------------SDSKALDLLDRMLT 312
sp|P53778|MK12_HUMAN      PAEFVQRLQSDEAKNYMKGLPELEKKDFASILTN------------ASPLAVNLLEKMLV 293
sp|O15264|MK13_HUMAN      GTEFVQKLNDKAAKSYIQSLPQTPRKDFTQLFPR------------ASPQAADLLEKMLE 290
sp|Q15759|MK11_HUMAN      SPEVLAKISSEHARTYIQSLPPMPQKDLSSIFRG------------ANPLAIDLLGRMLV 290
sp|Q16539|MK14_HUMAN      GAELLKKISSESARNYIQSLTQMPKMNFANVFIG------------ANPLAVDLLEKMLV 290
sp|P45983|MK08_HUMAN      CPEFMKKLQP-TVRTYVENRPKYAGYSFEKLFPDVLFPADSEHNKLKASQARDLLSKMLV 303
sp|P53779|MK10_HUMAN      CPEFMKKLQP-TVRNYVENRPKYAGLTFPKLFPDSLFPADSEHNKLKASQARDLLSKMLV 341
sp|P45984|MK09_HUMAN      SAEFMKKLQP-TVRNYVENRPKYPGIKFEELFPDWIFPSESERDKIKTSQARDLLSKMLV 303
sp|Q13164|MK07_HUMAN      SPAVIQAVGAERVRAYIQSLPPRQPVPWETVYPG------------ADRQALSLLGRMLR 329
sp|P31152|MK04_HUMAN      REEDKDELLR-VMPSFVS-STWEVKRPLRKLLPE------------VNSEAIDFLEKILT 294
sp|Q16659|MK06_HUMAN      HEEDRQELLS-VIPVYIRNDMTEPHKPLTQLLPG------------ISREALDFLEQILT 298
                                :      ::                 :                  * .:* ::*

sp|P28482|MK01_HUMAN      FNPHKRIEVEQALAHPYLEQYYDP-SDEPIAEAPFKFDMELDDLPKEKLKELIFEETAR- 353
sp|P27361|MK03_HUMAN      FNPNKRITVEEALAHPYLEQYYDP-TDEPVAEEPFTFAMELDDLPKERLKELIFQETAR- 370
sp|P53778|MK12_HUMAN      LDAEQRVTAGEALAHPYFESLHDT-EDEPQVQK-YDDSFDDVDRTLDEWKRVTYKEVLS- 350
sp|O15264|MK13_HUMAN      LDVDKRLTAAQALTHPFFEPFRDP-EEETEAQQPFDDSLEHEKLTVDEWKQHIYKEIVN- 348
sp|Q15759|MK11_HUMAN      LDSDQRVSAAEALAHAYFSQYHDP-EDEPEAEP-YDESVEAKERTLEEWKELTYQEVLS- 347
sp|Q16539|MK14_HUMAN      LDSDKRITAAQALAHAYFAQYHDP-DDEPVADP-YDQSFESRDLLIDEWKSLTYDEVIS- 347
sp|P45983|MK08_HUMAN      IDASKRISVDEALQHPYINVWYDPSEAEAPPPKIPDKQLDEREHTIEEWKELIYKEVMDL 363
sp|P53779|MK10_HUMAN      IDPAKRISVDDALQHPYINVWYDPAEVEAPPPQIYDKQLDEREHTIEEWKELIYKEVMNS 401
sp|P45984|MK09_HUMAN      IDPDKRISVDEALRHPYITVWYDPAEAEAPPPQIYDAQLEEREHAIEEWKELIYKEVMDW 363
sp|Q13164|MK07_HUMAN      FEPSARISAAAALRHPFLAKYHDP-DDEPDCAPPFDFAFDREALTRERIKEAIVAEIEDF 388
sp|P31152|MK04_HUMAN      FNPMDRLTAEMGLQHPYMSPYSCP-EDEPTSQHPFRIEDEIDDIVLMAANQSQLSNWDTC 353
sp|Q16659|MK06_HUMAN      FSPMDRLTAEEALSHPYMSIYSFP-MDEPISSHPFHIEDEVDDILLMDETHSHIYNWE-- 355
                          :.   *: .  .* *.::     .   *.            :          .     :
```

Fig. 4 CLUSTALW multiple sequence alignments of MAPK 1–14 goes here

Table 1 Active site residue variations among selected MAPKs

MAPK 3GC8 active site residues	1	3	4	6	7	8	9	10	11	12	13	14
Tyr35		N	N	N		Q	Q	Q				
Ser154		E	A	A						G	G	
Val30	I	L	L	L	I	I	I	I				
Ala111	T	G	T	T						T	T	
Gly110	E	E	E	E	E	D	D	D			Q	
Leu108	I	I	I	I	V	I						
Met109												
Leu167	C	T	G	G	G							
Ala51	*	*	*	*	*	*	*	*	*	*	*	*
Lys53	*	*	*	*	*	*	*	*	*	*	*	*
Leu104	I	I	I	I	V	I	I					
Tyr106	Q	M	Q	Q	L	M	M	M		M	M	

* represent conserved sites through all MAPKs. Amino acid variations are represented in single letter codes. Empty cells represent 3GC8 identical residues

3.1 Phylogeny

From the phylogenetic tree, it is evidenced that all MAP Kinases 1 to 13 in humans participate in different pathways. MAPK 1 and 7 are on one taxon and role of MAPK 7 is unknown but MAPK1 is involved in growth factor pathway. MAPK 2, 3 and 5 are on one taxon. MAPK3 is involved in growth factor pathway and remaining two MAP Kinases do not participate in any of the signalling pathways of the MAPKs. MAPK 4 and 6 is on one taxon. These two kinases also do not not participate in any of the signalling pathways. MAPK 8, 9 and 10 are on one taxon. These three are involved in stress pathway. MAPK 11, 12 and 13 are on one taxon and these three are involved in inflammatory pathway.

3.2 Identification of Phosphorylation Sites Presented in Tables

All MAPKs are investigated for the presence of similarities of specific amino acid residues at the active site regions. Not all MAPKs have 3-dimensional structure and hence comparison of active site residues were made with known 3-D structures using multiple alignments. Table 9 presents the information about MAP Kinases participating in stress, growth and inflammatory pathways and its PDB structures (Table 10).

The present work reported the identification of pathway specificity of MAPK's participating in the MAPK signaling pathway. The sequence of all the MAPK's which participate in the pathway are extracted from EXPASY (Expert protein analysis system). Around 13 sequences of humans are obtained and are subjected to Sequence Analysis and Phylogenetic Analysis by using CLUSTALW. It has been observed from the analyses that the path specific MAPK's share a very close proximity on the phylogenetic trees by originating from a common node and diverging to a very small extent. Finally the path specificity of the MAPK sequences is confirmed by applying the results of the above analyses for better understanding. Active sites analysis performed on MAPK's, it was observed that the main driving force of the pathway specificity was due to the presence of specific domains in protein sequences and the MAPKs which participate in the same pathway have been hypothesized to have more or less similar domains in their protein sequence.

MAPK1:

Table 2 Phosphorylation sites of MAPK1

Domains	Begining	Ending	Phosphorylated sites
PDA1T9W2	11	30	Serine-28
PD972663	19	100	Threonine-62
PD000001	31	230	Threonine 180, 184, 189 Tyrosine 186 Serine-201
PD696917	277	312	Serine-283

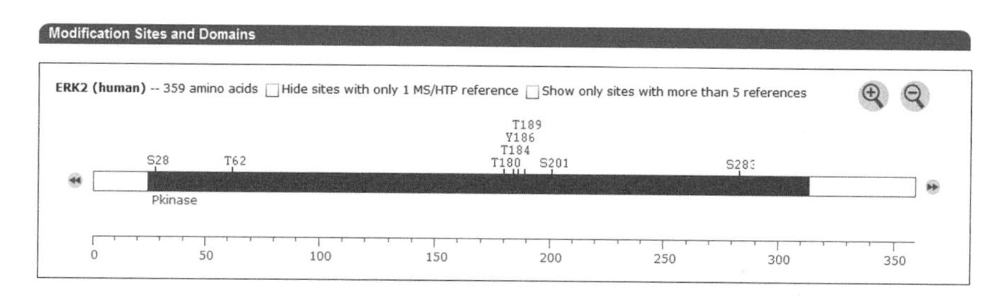

Fig. 5 Phosphorylation sites in MAPK1 on all domains goes here

MAPK3:

Table 3 Phosphorylation sites of MAPK3

Domains	Begining	Ending	Phosphorylation sites
PD000001	43	303	Threonine-201 Tyrosine-204, 207, 208 Serine 251
PDA1I9T1	304	332	Threonine-313, 317

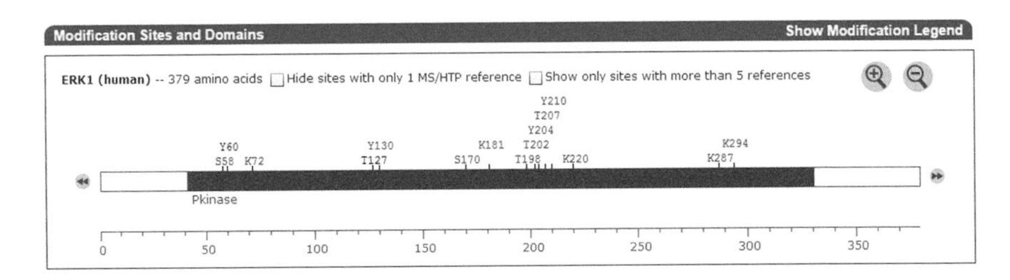

Fig. 6 Phosphorylation sites in MAPK3 on all domains goes here

3.2.1 Stress Related Pathway

MAPK8:

Table 4 Phosphorylation sites of MAPK8

Domains	Begining	Ending	Phosphorylated sites
PD067788	10	171	Serine-155
PD000001	29	274	Theronine-178, 183, 188 Tyrosine-185

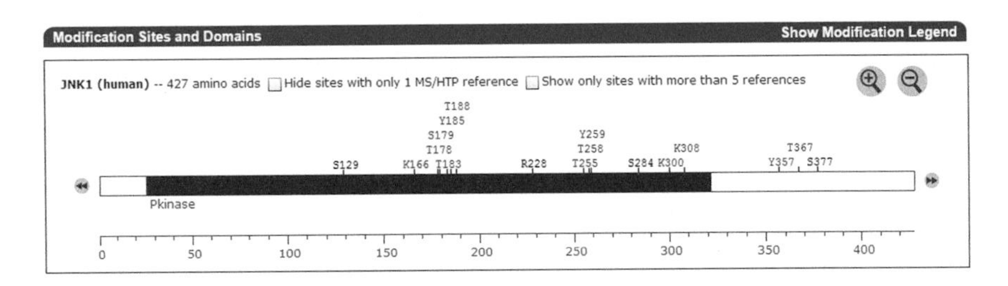

Fig. 7 Phosphorylation sites in MAPK8 on all domains goes here

MAPK9:

Table 5 Phosphorylation sites of MAPK9

Domains	Begining	Ending	Phosphorylated sites
PD067788	8	227	Serine-155 Theronine-183 Tyrosine-185

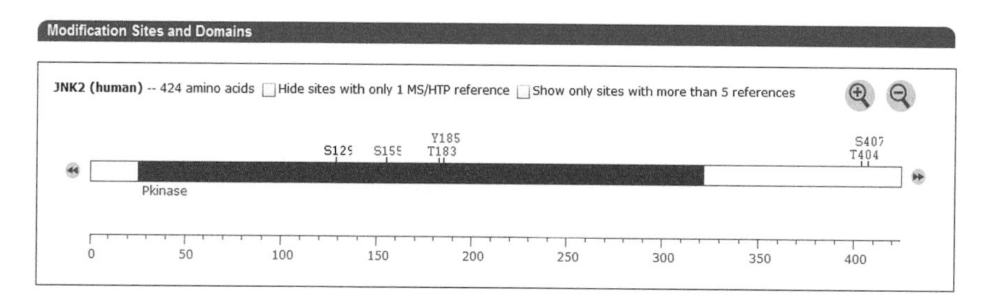

Fig. 8 Phosphorylation sites in MAPK9 on PD067788 domain goes here

MAPK10:

Table 6 Phosphorylation sites of MAPK10

Domains	Begining	Ending	Phosphorylated sites
PD067788	48	209	Serine-193

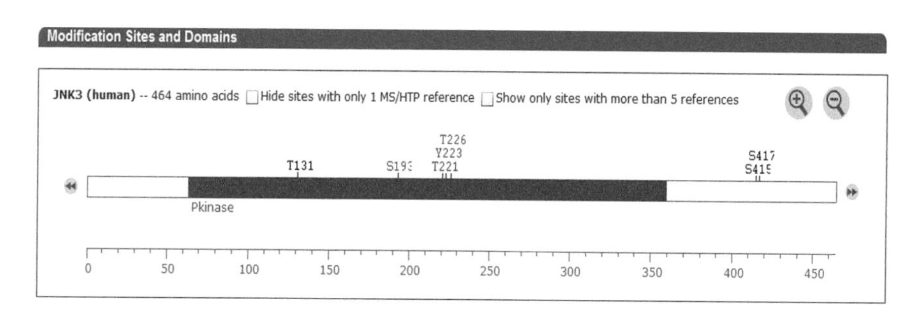

Fig. 9 Phosphorylation sites in MAPK10 on PD067788 domain goes here

3.2.2 Inflammatory Pathway

MAPK11:

Table 7 Phosphorylation sites of MAPK11

Domains	Begining	Ending	Phosphorylated sites
PD067788	20	209	Tyrosine-182
PD000001	27	243	Serine-243

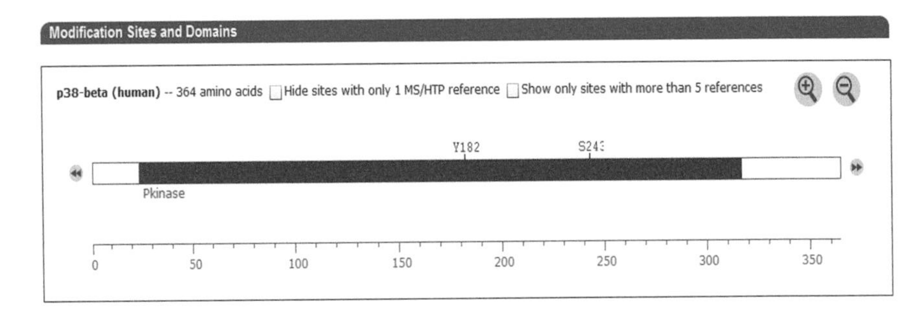

Fig. 10 Phosphorylation sites in MAPK11 on PD067788 and PD000001 domains goes here

MAPK12:

Table 8 Phosphorylation sites of MAPK12

Domains	Begining	Ending	Phosphorylated sites
PDA1T9N7	1	26	Serine-3
PD000001	27	265	Theronine-183 Tyrosine-185

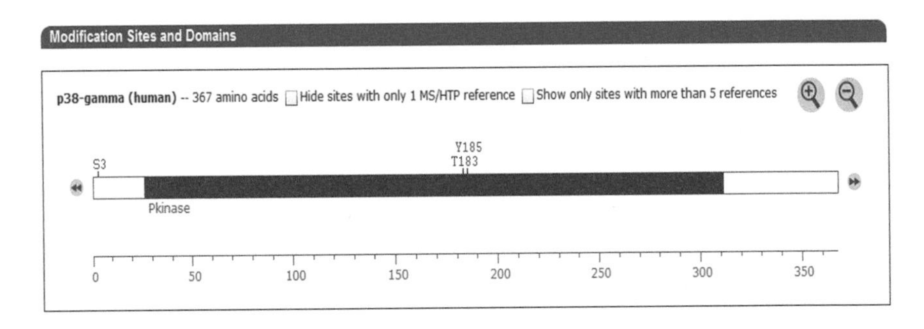

Fig. 11 Phosphorylation sites in MAPK12 on PDA1T9N7 and PD000001 domains goes here

MAPK13:

Table 9 Phosphorylation sites of MAPK13

Domains	Begining	Ending	Phosphorylated sites
PD000001	32	237	Serine-47 Tyrosine-182

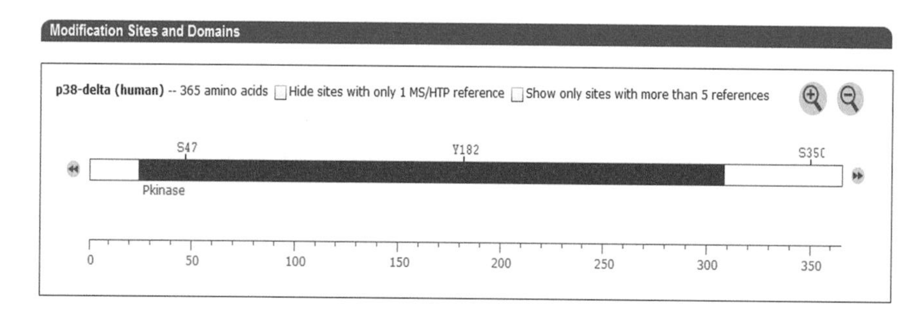

Fig. 12 Phosphorylation sites in MAPK13 on PD000001 domain goes here

Table 10 MAPK kinases with their uniprot id and PDB entries

S no	Mapkinases	Uniprot ID	Pathway	PDB structure
3	MAPK 3	Q16644	Growth	2Z0Q
4	MAPK 4	P31152	Unknown	Undetermined
6	MAPK 6	Q16659	Unknown	216L
7	MAPK 7	Q13164	Unknown	28QY
8	MAPK 8	P45983	Stress	1UKI
9	MAPK 9	P45984	Stress	3E70
10	MAPK 10	P53779	Stress	1JNK
11	MAPK 11	Q15759	Inflammatory	3GC8
12	MAPK 12	P53778	Inflammatory	1CM8
13	MAPK 13	015264	Inflammatory	3C01
14	MAPK 14	Q16539	Inflammatory	1W82

4 Conclusions

The specificity of each MAPK specific to a particular pathway was studied from the Phylogenetic Tree Analysis and identification of phosphorylated sites. Computational analysis provided identification of specific MAPK participating in a particular pathway and phylogenetic studies revealed that the path specific MAPK's show close proximity in the phylogenetic tree, originating from a common node and diverging to a very extent, thus showing greater degree of similarity among them. Phosphorylated sites revealed similar and dissimilar domains in MAPKs. Amino acid variations among MAPKs suggest that specificity in protein sequence conservation exists among all MAP kinases and those with slight variations participate in defined pathways. Structural data comparison of similar 3GC8 active site residues with other proteins showed that MAPK 12 and 13 diverged from MAPK 11 and 14.

References

1. Lewis TS, Shapiro PS, Ahn NG (1998) Signal transduction through MAP kinase cascades. Adv Cancer Res 74:49–139
2. Errede B, Cade RM, Yashar BM, Kamada Y, Levin DE, Irie K, Matsumoto K (1995) Dynamics and organization of MAP kinase signal pathways. Mol Reprod Dev 42:477–485
3. Kyriakis JM, Avruch J (2012) Mammalian MAPK signal transduction pathways activated by stress and inflammation: a 10-year update. Physiol Rev 92(2):689–737
4. Kultz D (1998) Phylogenetic and functional classification of mitogen and stress-activated protein kinases. J Mol Evol 46:571–588
5. Fu M, Wang C, Zhang X, Pestell RG (2004) Signal transduction inhibitors in cellular function. Methods Mol Biol 284:15–36
6. Cuevas BD, Abell AN, Johnson GL (2007) Role of mitogen-activated protein kinase kinase kinases in signal integration. Oncogene 26:3159–3171
7. Cobb MH, Goldsmith E (1995) How MAP kinases are regulated. J Biol Chem 270:14843–14846
8. Sabapathy K, Hu Y, Kallunki T, Schreiber M, David JP, Jochum W, Wagner EF, Karin M (1999) JNK2 is required for efficient T-cell activation and apoptosis but not for normal lymphocyte development. Curr Biol 9:116–125
9. Rouse J, Cohen P, Trigon S, Morange M, Alonso-Llamazares A, Zamanillo D, Hunt T, Nebreda AR (1994) A novel kinase cascade triggered by stress and heat shock that stimulates MAPKAP kinase-2 and phosphorylation of the small heat shock proteins. Cell 78(6):1027–1037
10. Chang L, Karin M (2001) Mammalian MAP kinase signalling cascades. Nature 410(6824):37–40
11. Ubersax JA, Ferrell JE Jr (2007) Mechanisms of specificity in protein phosphorylation. Nat Rev Mol Cell Biol 8:530–541. doi:10.1038/nrm2203
12. Aberg E et al (2006) Regulation of MAPK-activated protein kinase 5 activity and subcellular localization by the atypical MAPK ERK4/MAPK4. J Biol Chem 281:35499–35510
13. Patel SB et al (2009) The three-dimensional structure of MAP kinase p38beta: different features of the ATP-binding site in p38beta compared with p38alpha. Acta Crystallogr D Biol Crystallogr 65:777–785
14. Zhang W, Liu HT (2002) MAPK signal pathways in the regulation of cell proliferation in mammalian cells. Cell Res 12:9–18
15. Seger R, Krebs EG (1995) The MAPK signaling cascade. FASEB J 9:726–735
16. www.biocarta.com/

Identification of Drug Targets from Integrated Database of Diabetes Mellitus Genes Using Protein-Protein Interactions

Duggineni Kalyani, Naresh Babu Muppalaneni, Ch Ambedkar and Kiran Kumar Reddi

Abstract In this paper, we looked at clustering coefficient, one of the statistics commonly used to study protein interaction networks. We have calculated clustering coefficient to identify the potential drug targets for Diabetes Mellitus (DM). An integrated database of DM genes has been developed by mining various repositories. Protein-protein interaction (PPI) network for the DM genes was constructed. Various centrality measures for the PPI network were calculated. In this study we considered the top 10 genes from the 15 centrality measures and further we calculated the clustering coefficient to find the potential drug targets.

1 Introduction

Diabetes mellitus is a metabolic disorder which is characterized by resistance to insulin and abnormalities in the hepatic cells in terms of glucose production [1]. It has been known to be a complex and multi factorial disease as it affects various other organ systems in the body like cardiovascular system, kidneys, lungs, eyes etc. [2–4].

Many online databases are available which consists of DM genes. But candidate genes identification is crucial. Drug-development strategies based on target-driven approaches is an efficient method to combat a certain disease was sought [5–7]. We have carried out the study to investigate the drug targets for Diabetes Mellitus.

D. Kalyani (✉) · N.B. Muppalaneni
C R Rao AIMSCS, Hyderabad, India
e-mail: kalyani.duggineni@gmail.com

Ch. Ambedkar
SRK College of Engineering, Vijayawada, India

K.K. Reddi
Krishna University, Machilipatnam, India

R. Bhramaramba and A.C. Sekhar (eds.), *Application of Computational Intelligence to Biology*, Springer Briefs in Forensic and Medical Bioinformatics,
DOI 10.1007/978-981-10-0391-2_8

With the increasing of human protein interaction data, the entanglement of the techniques can be vanquished through protein–protein interaction networks (PPINs) [8, 9].

Protein-protein interaction maps provide a valuable framework for a better understanding of the functional organization of the proteome. PPI plays important role in investigating the protein function of target protein and drug ability of molecules [10–13].

2 Methodology

The network approach examines the efficacy of drugs in the context of a network of relevant PPI [5–7]. In the protein-protein interaction network, each protein is a vertex and the edge represents the interaction between two proteins. The network approach now has a tradition in drug target analysis.

2.1 Data Set

We have developed an integrated database for the human genes involved in Diabetes mellitus from various sources like MalaCards, DMBase and Jensen Group (Jensenlab) of Novo Nordisk Foundation Center for Protein Research, Denmark, which maintains DISEASES database. DISEASES database is a frequently updated web resource that integrates evidence on disease-gene associations from automatic text mining, manually curated literature, and genome-wide association studies. We got that 1220 genes causing diabetes mellitus, after eliminating the duplicate entries and unknown HPRD ids it reduced to 1085 genes.

2.2 PPI

Typical examples for networks which can be analyzed are protein-protein interaction networks. All interactions between proteins of an organism can be represented as a network. Several databases contain information about such protein-protein interactions.

Protein-protein interactions for 1085 DM genes are extracted from Human Protein Reference Database (HPRD) [14, 15].

2.3 Network Properties

Centrality measures are indicators of the importance of a node in a network. We find out different graph centrality measures such as degree, eccentricity, closeness, centroid values, shortest-path betweenness, current-flow closeness, current-flow betweenness, Katz status index, Eigen vector for every node in the interactome and rank them according to their scores which further aids in establishing the properties of protein interaction network.

Clustering Coefficient

To gain more understanding on the probability of proteins being potential drug targets, we developed a model to quantify the clustering of drug targets, disease genes, and essential genes surrounding other proteins.

Clustering coefficient of a node is defined as, Let a node S of a graph has N neighbors. The maximum possible number of edges among the N neighbors is N*(N−1)/2. The fraction of this maximum possible number of edges that actually exist between neighbors of S is its clustering coefficient. The value of clustering coefficient is between zero and one that is zero when there is no clustering, and one for maximal clustering.

3 Results

We have constructed the Human protein-protein interaction network (HPPI) from the data collected from the HPRD database. Out of 1085 proteins we got HPRD IDs for 1066 proteins.

After curation of the data sets, removal of self edges and parallel edges the Human PPI network comprises of 1027 vertices with 2950 edges. We have

Table 1 Top 10 genes of 15 centrality measures with their clustering coefficient

Gene symbol	Cluster coefficient
SRC	0.074125874
ESR1	0.108123904
MAPK1	0.078644888
AKT1	0.065993266
GRB2	0.093333333
EGFR	0.086530612
STAT3	0.139372822
MAPK3	0.06763285
PIK3R1	0.129292929
TP53	0.051428571
HSP90AA1	0.115864528
FYN	0.077777778
APP	0.018115942
PTPN11	0.205705706

calculated fifteen different graph centrality measures for the PPI network constructed and we calculated the clustering coefficient for top 10 genes of centrality measures. The top 10 genes from 15 centrality measures with their clustering coefficient are presented in Table 1. Among all these 14 genes 5 genes showing high clustering coefficient obtained are as follows ESR1, STAT3, PIK3R1, PTPN11 and JAK2. Although JAK2 is not included in the set of these 14 genes, it was found to be showing high clustering coefficient (0.2235), predicted to be the potential drug target for Diabetes mellitus.

4 Conclusion

Our work prioritizes the importance of applications graph theoretical approaches on biological networks by identification of drug target in human protein- protein interaction network through network analysis. This helped to identify genes that are highly critical in DM. ESR1, STAT3, PIK3R1, PTPN11 and JAK2 are considered to be potential drug targets for DM.

Acknowledgments Dr. M. Naresh Babu and Ms. Duggineni Kalyani would like to thank SERB, Department of Science and Technology, Government of India under FAST Track scheme order No. SB/FTP/ETA-436/2012.

References

1. Fujimoto WY (2000) The importance of insulin resistance in the pathogenesis of type 2 diabetes mellitus. Am J Med 108:9–14
2. McIntyre EA, Walker M (2002) Genetics of type 2 diabetes and insulin resistance: knowledge from human studies. Clin Endocrinal 57:303–311
3. Tusie LMT (2005) Genes and type 2 diabetes mellitus. Arch Med Res 36:210–222
4. Kaluga M (2006) Insulin resistance and pancreatic beta cell failure. J Clin Invest 116:1756–1760
5. Watts DJ, Strogatz SH (1998) Collective dynamics of 'small-world' networks. Nature 393:440–442
6. Barabási AL, Albert R (1999) Emergence of scaling in random networks. Science 286:509–512
7. Newman MEJ (2003) The structure and function of complex networks. SIAM Rev 45:167–256
8. Barabási A-L, Gulbahce N, Loscalzo J (2011) Network medicine: a network-based approach to human disease. Nat Rev Genet 12:56–68
9. Wang E, Zou J, Zaman N, Beitel LK, Trifiro M, Paliouras M (2013) Cancer systems biology in the genome sequencing era: part 1, dissecting and modeling of tumor clones and their networks. Semin Cancer Biol 23:279–285
10. Albert R (2002) Statistical mechanics of complex networks, Rev Mod Phys 74:47–97
11. Milo R et al (2002) Network motifs: simple building blocks of complex networks, Science 298:824 [PMID: 12399590]
12. Holme P et al (2003) Subnetwork hierarchies of biochemical pathways, Bioinformatics 19:532 [PMID: 12611809]
13. Wuchty S et al (2003) Centers of complex networks, J Theor Biol 223:45 [PMID: 12782116]
14. Prasad, TS Keshava, et al (2009) Human protein reference database—2009 update. Nucleic Acids Res 37:D767–D772
15. http://hprd.org

Distributed Data Mining for Modeling and Prediction of Skin Condition in Cosmetic Industry—A Rough Set Theory Approach

P.M. Prasuna, Y. Ramadevi and A. Vinay Babu

Abstract Cosmetic industry is proliferating rapidly these days expanding its business globally with spatial distribution. However to rejuvenate a product to deal with a specific problem, analyzing data at the local level is not sufficient as the influencing factors of facial skin issues may vary from region to region. This leads to the situation where one needs to analyze the data in distributed environment in which local models are merged and further mined at the central node to derive the global modal which gives the adequate information to have better understanding of the skin problem thereby helping the industry to know what are the most common problems the people are suffering from and what type of products they are expecting from the industry. This paper discusses how to mine the cosmetic data in distributed environment using rough set theory.

Keywords Distributed data mining · Rough set theory · Local modal · Global modal · Discretization · Mean roughness · Distance relevance · Maximum attributes dependency

1 Introduction

Cosmetic industry is a huge industry with abundant data of its customers. It is a globally wide spread organization collecting the facial skin features of millions of customers. This data provides the industry to have a better understanding of the

P.M. Prasuna (✉)
JNTU, Hyderabad, India
e-mail: prasunamanikya@yahoo.com

Y. Ramadevi
CBIT, Hyderabad, India
e-mail: 2yrdcse.cbit@gmail.com

A. Vinay Babu
JNTUHCE, Hyderabad, India
e-mail: 3avb1222@jntuh.ac.in

R. Bhramaramba and A.C. Sekhar (eds.), *Application of Computational Intelligence to Biology*, Springer Briefs in Forensic and Medical Bioinformatics,
DOI 10.1007/978-981-10-0391-2_9

facial skin related issues of the customers spread worldwide. The issues depend on many factors like environmental conditions in which customers live in, the food the customers take, their working hours, sleeping hours and nature of the jeans they adopted from their ancestors [1]. These factors may vary from region to region. Data collected from a particular part may not give sufficient information to come to a concrete conclusion about suitable product to deal with the skin issues, the customers are suffering from. Data has to be collected from different regions and store at a center repository and analyze further. Hence it is required to gather the data from different parts and have a clear insight of the data to analyze the skin problems. Good analysis of this data leads to rejuvenate a cosmetic product that deals with the problem in an effective way.

In this regard the industry has to have insight knowledge of the skin conditions of its customers from various regions of the world. First they will capture the face images of the customers in a very sophisticated environment. Various features like skin color, hydration level, wrinkles, acne count, spot count are then extracted from the images. These features are further used to analyze and discover the skin condition of the customer.

Hence it is evident that analyzing the data, identifying the issues related with facial skin and either suggesting a product or rejuvenating a product needs some mining technology to be applied in distributed environment [2]. The proposed method first preprocesses the data. The preprocessing is required as data extracted from the facial images is numeric. Many mining algorithms work on categorical data and he proposed rough set theory approach also take only discretized data s the input. More over the final rules derived will be precise and more accurate only when discretized data is used. Hence the data is first discretized using rough set theory approach. Not all the features extracted from the facial images may play important role in the final model to be derived. The irrelevant features are identified and removed, thereby generating final reduct. Clusters are generated considering only the features contained in the reduct using rough clustering technique. These clusters are then transformed into classes by identifying the most dominant feature in each cluster and labeling it as the class label for the respective cluster. The classes are then used to derive the rules.

But these rules can only be used for knowing the skin conditions of customers of a particular region only. To get the global result the entire process has to be applied in distributed environment. First the analysis is done at local level i.e. capturing face images, extracting the features, reduct generation, formation of clusters, transformation of clusters into classes then finally discovering rule are done at individual nodes. The rules thus discovered are then sent to the central node where the global model is derived basing on the rules received from various local sites/nodes.

This paper has been organized as follows. Section 2 gives the brief introduction about distributed data mining, Sect. 3 presents the concepts of Rough Set Theory, Sect. 4 deals with preprocessing step, Sect. 5 discusses rough clustering, reduct finding, Sect. 6 deals with rule discovery and Sect. 7 and 8 presents results and conclusion.

2 Introduction to Distributed Data Mining

In today's business field organizations are running their business at different parts of the world. They are managing and maintaining their databases at different locations and can say distributed globally. To extract the unified knowledge from these distributed data bases either homogeneous or heterogeneous there is a need to have some mining algorithms. As these mining algorithms have to deal with voluminous data located at different places the communication cost is extensively high. Moreover the traditional data mining algorithms for geographically spread users and data would be insufficient to deal with distributed environment. The development of data mining along this dimension has lead to the emergence of distributed data mining. Distributed Data Mining (DDM) is a branch of the field of data mining that offers a framework to mine distributed data paying careful attention to the distributed data and computing resources. At the same time there is a need to address many issues like data, hardware and the mining software associated with the application of data mining in distributed computing environment.

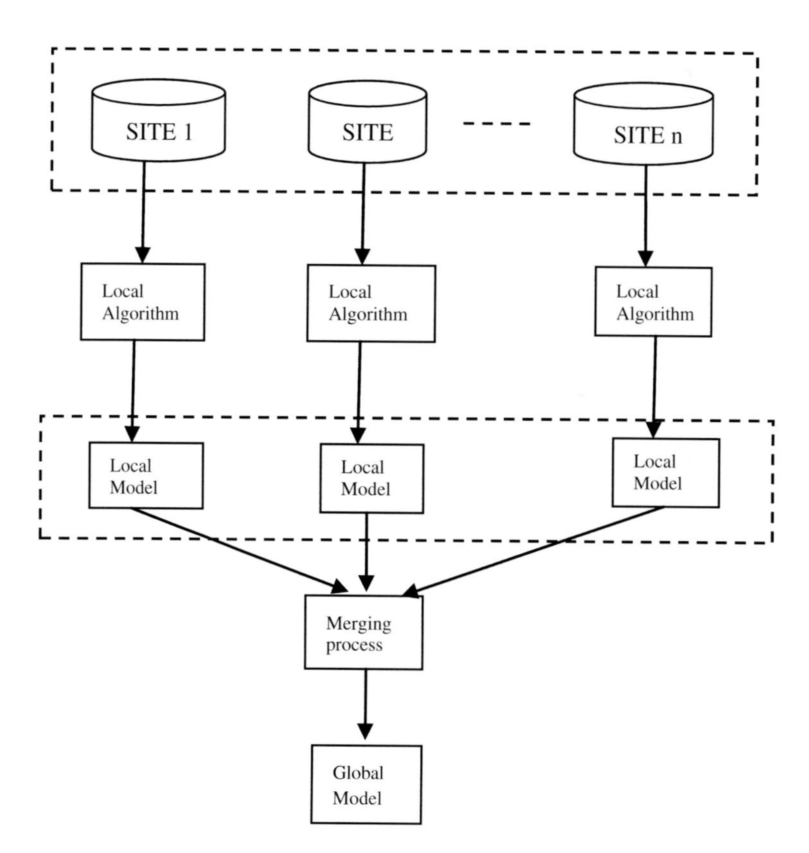

Fig. 1 Distributed data mining

The Mining algorithms can be applied to each distributed site and as well as at a global level to which the locally generated knowledge is merged to derive the global knowledge. The architecture of a Distributed Data Mining approach is shown in Fig. 1. Phase 1 handles the analysis of the local database at each distributed site. Then, the derived knowledge is transmitted to a central site, where the integration of the distributed local models is performed. Once the global knowledge is discovered, the local sites are updated with this discovered knowledge. In some cases, the local sites work in parallel and share the knowledge to derive the global model. Distributed databases may have homogeneous or heterogeneous schemata. In the former case, the attributes describing the data are the same in each distributed database. This is often the case when the databases belong to the same organization (e.g. local stores of a chain). In the latter case the attributes differ among the distributed databases. In certain applications a key attribute might be present in the heterogeneous databases, which will allow the association between tuples.

3 Rough Set Theory

The data we deal with everyday is sometimes uncertain. The available data may be inadequate to derive the useful knowledge. This vague knowledge can be handled efficiently using Rough set theory approach [3]. Pawlak introduced the Rough set theory in 1985. It is mainly used to disclose hidden patterns from imperfect data. When the available information is incomplete or we can say insufficient to yield the results, the concepts lower and upper approximations in the Rough set theory are helpful. Rough set theory is nowadays being used widely to solve problems in various areas. The concepts involved in Rough set theory are presented below.

3.1 Information/Decision Systems (Tables)

An information system or information table also called decision table consists of objects (rows) and attributes (columns). More formally, the information system IS is represented as IS = (U, A, V, f), where: U is a finite sets of objectives (universe), A = {a1, a2, …, am} is a finite set of attributes, V_a is the domain of the attribute a, $V = U_{a} \in AV_a$ and f: $U \times A \rightarrow V$ is a total function such that $f(x, a) \in V_a$ for each $a \in A$, $x \in U$, called information function.

3.2 Indiscernibility

Indiscernibility Relation is the main concept in Rough Set Theory. In an information table, if a relation gives identical values for the subset of attributes for two or more objects then the relation is indiscernibility relation. For every set of attributes P: A, if two objects, x_i and x_j, are indiscernible by the set of attributes P in A, if b(xi) = b(xj) for every p ϵ P then the relation between the objects is known as indiscernibility relation and is denoted by IND(P). Since the equivalence class of this indiscernibility relation an indiscernibility relation represents smallest discernible groups of objects, it is called as elementary set in [4]. For any element xi of U, the equivalence class of xi in relation IND(P) is represented as [xi]IND(P).

3.3 Set Approximation

The important objective of the Rough set theory is defining the lower and the upper approximations of a set. As Rough set theory is widely used when the data is imperfect or incomplete, in the data analysis, the first step is to identify the objects which certainly belong to the set and identify objects which may belong to the set and is done by determining the lower and upper approximations of the set respectively. Let X denote the subset of elements of the universe U (X ⊂ U).

The lower approximation of X in B(B ⊂ A) is denoted as BX, is defined as the union of all these elementary sets which are contained in X.

$$\underline{BX} = \{x_i \in U | [x_i]_{Ind(B)} \subset \mathrm{X}\}$$

The upper approximation of the set X, denoted as BX, is the union of these elementary sets, which have a non-empty intersection with X:

$$\overline{BX} = \{x_i \in U | [x_i]_{Ind(B)} \cap X \neq 0\}$$

B-boundary region of X, $BN_B(X) = \overline{BX} - \underline{BX}$ consists of those objects that we cannot decisively classify into X in B. If the lower and upper approximation are identical i.e., $\underline{BX} = \overline{BX}$, then set X is definable, otherwise, set X is undefinable in U [5]. There are four types of undefinable sets in U:

1. if $\underline{BX} \neq \emptyset$ and $\underline{BX} \neq U$, X is called roughly definable in U.
2. if $\underline{BX} \neq \emptyset$ and $\underline{BX} = U$, X is called externally undefinable in U.
3. if $\underline{BX} = \emptyset$ and $\underline{BX} \neq U$, X is called internally undefinable in U.
4. if $\underline{BX} = \emptyset$ and $\underline{BX} = U$, X is called totally undefinable in U, where Ø denotes an empty set.

3.4 Accuracy of Approximations

$$\propto_B (X) = \frac{\underline{BX}}{\overline{BX}}$$

where S = (U, A), B ⊆ A and X ⊆ U |X| denotes the cardinality of X If αB(X) = 1, then X is crisp with respect to B. If αB(X) < 1, then X is rough with respect to B.

3.5 Rough Membership

The rough membership function is used to determine the probability that x belongs to X given R. In other terms we can say rough membership functions gives the degree that x belongs to X in view of information about x expressed by R [6]. The lower and upper approximations and also the boundary region can be defined by the rough membership function. When one considers subsets of a given universe it is possible to apply characteristic functions for expressing the fact whether or not a given element belongs to a given set.

The rough membership function of the set X denoted by μAX, is defined as follows:

$$\mu X^{A}(X) = \frac{|[x]_{A} \cap X|}{|[x]_{\mathrm{A}}|} \quad \text{for x} \in \text{U}$$

3.6 Dependency of Attributes

Most of the times the data we take for analysis, contain irrelevant features/attributes. These irrelevant features are to be eliminated, retaining only the features/attributes which play important role in data analysis. The attributes which have very negligible influence on other attributes are identified by measuring degree of dependency among the attributes. Dependency of attributes can be measured using Rough set theory as follows Let *A* and P be subsets of attribute set, we say that P depends on A in a degree k ($0 \leqq k \leqq 1$)

$$k = \gamma(A, P) = \sum_{X \in U/D} \frac{|\underline{A}(X)|}{|U|}$$

$$\mathrm{k} = \gamma(\mathrm{A}, \mathrm{P}) = \frac{|\mathrm{POS_A P}|}{|\mathrm{U}|}$$

where A and P belong to the attribute set. $|\mathrm{POS_A P}|$ is the A-positive region of P. The value k determines the dependency of P on A. I k is less than 1 then attribute set P depends on attribute set A partially. If k is 1 then attribute set P depends totally on attribute set A.

4 Preprocessing

As the proposed procedures uses Rough set theory and Rough Set theory can only be applied on categorical data and the features extracted from face images of customers is continuous the first step in the mining process is discretization. The discretization process in distributed environment goes as follows:

1. First discretization is done at the local nodes using the proposed algorithm given below.
2. The cluster centroids are sent to the central node as data set.
3. At the central node clusters are formed on the data set this again using the same algorithm used at local nodes.
4. The new generated centroids are sent to the local node.
5. Assignment of each object to new centroid closer to it sent by global modal is done at each local node which gives the exact cut points to discretize data at local nodes.

This discretization process both at local site and central site is carried out into two phases-in phase 1, traditional K-means algorithm is applied to form clusters from the given unsupervised data. In phase 2 these clusters are further refined using Rough set approach to make objects with similar traits fall into one cluster which decide the right cut points to discretize the data. The discretization process is carried in distributed environment as follows.

Phase 1:

K-means clustering technique is an unsupervised learning algorithm to generate clusters. The procedure first decides number of clusters to be formed (assume k clusters). Then k centroids are defined, which is selecting k data points randomly. Then it takes each point belonging to the data set and associates to the nearest centroid. When all data points are attached to the nearest centroid, new centroids are computed. After we have these k new centroids, a new binding has to be done between the same data set points and the nearest new centroid. This is a repetitive process and this process is repeated until no more changes are done. This algorithm also deals with minimizing the objective function. In this algorithm a squared error function [7] is used which is

$$J = \sum_{j-1}^{k} \sum_{i-1}^{n} \left\| x_i^{(j)} - c_j^2 \right\|$$

where $\left\| x_i^{(j)} - c_j^2 \right\|$ is distance between a data point $x_i^{(j)}$ and the cluster centre c_j, gives distance of the n data points from their respective cluster centers.

The algorithm consists of the following steps:

1. Initially select K points which represent the objects randomly.
2. These points are the centroids for k clusters initially.
3. Determine the distance between each object to the k centroids and assign the point to the centroid closer to it.
4. Repeat step 3 until there are no more objects to be assigned.
5. Compute the new K centroids.
6. Repeat Steps 3–5 until the centroids no longer change.

Phase 2:

Since the k means algorithm suffers with limitations like Handling Empty Clusters, Outliers and Difficulty in measuring the no of clusters, the results obtained from this algorithm may not be accurate. Hence discretization using clustering technique is not sufficient to generate cut points with minimum information loss [8]. Hence they are refined using Rough set theory concepts. The refinement process consists of either merging or splitting of clusters. This process will give better clustering by reducing the sum of squared error. The refinement is to enhance the significance of the attribute. In rough set theory the significance of an attribute is measured through rough membership function POSai(D). Thus maximizing POSai(D) leads to maximizing the significance of the attribute. To maximize POSai(D), the clusters formed through kmeans are refined further to generate new intervals or cut points.

Let us consider data set U having n clusters say {k1, k2, k3 … kn}. The rough membership function for any attribute a_i with sorted m distinct values as $\{v_{i1}, v_{i2}, v_{i3} \ldots v_{im}\}$ is calculated as follows.

Let us take any cluster, for example k_p. Consider the interval $I = [V_{i1}, V_{ij}]$ of attribute a_i for class k_p. Then the Rough membership function is

$$f(ai, k_p, I) = \frac{card(\underline{a_{i,I}} X_{kp})}{card(X_{a_{i,I}})}$$

where $X_{a_{i,I}} = \{x|ai(x)| \in I\}$ and, $\underline{a_{i,I}} X_{kp} = \left\{x|ai(x) \in I, D(x) = k_p\right\}$

Maximizing f(ai, k_p, I) is maximizing $card(\underline{a_{i,I}} X_{kp})$ which further maximizes POSai(D). To achieve this each cluster generated by kmeans is examined carefully and if necessary a cluster may be split into two or merged with the neighboring cluster [9]. The splitting process uses the rough set membership function such that it maximizes the POSai(D). In this way the intervals are refined. The refinement takes place as follows. Initially three predetermined parameters are taken. Max_size determines the maximum no of values that could fall in each cluster. Min_size decides the minimum number of values to form a cluster and Range gives the length of the cluster. These parameters decide whether the cluster can be retained or still to be refined. The refinement process takes place if the cluster is large or small. The cluster is said to be large if its cardinality is greater than the Max_size or the length is greater than the Range [10]. A cluster is treated as small if its cardinality is less than the Min_size. If the cluster is large it is split into two or else small, merged with other small clusters thereby generating new cut points or intervals. This process is refined until there is no change in the cut points or intervals [11].

Algorithm for the proposed method:

Step1: Consider each attribute in the data set, select distinct values and sort them.

Step2: Apply kmeans algorithm to form clusters.

Step3.From the generated clusters determine the class labels as well as intervals.

Step4.Refine the intervals and add new intervals to the interval set.

```
Refine (I1, I2, .... Ir))

While (no change in no. of intervals) do
For each interval Ij
If SP-C (Ij, Min_size, Range) = True then

        Temp= Cut Point ({vj1, vj2, vj3 ... vjk})
        Replace the interval Ij with two intervals
        Ij1 = [vj1, Temp] and Ij2 = [Temp, vjk]
Else if | Ij | < Min_size then
        If for Ik' either neighbour of Ij
        MR_C (Ij, Ik', Max_size, Min_size) = True   then Merge Ij to an interval Ik'

         End if
         End if

      End for
      End while


Cut Point ({vi1, vi2, vi3 ... vik})

I = [vi1, vik/2]
MAXRMV= Max ({f (Ai, cp, I)}) ∀ cp,
for each vij , j=k/2 to 2
      I = [vi1, vij]
      Temp = Max ({ f (A i , c p , I) }) ϵ cp;

      if Temp> MAXRMV then
             MAXRMV=Temp; else
             break;
        If (j < k/2) then return vij as cut point for the cluster
        else
    for each vij, j=k/2 to k-1
          I = [vij, vij]
          Temp = Max ({ f (A i , cp , I) }) ϵ cp;
          if Temp> MAXRMV then
                   MAXRMV=Temp;
              else
          return vij as cut point for the cluster
          else
          for each vij, j=k/2 to k-1
             I = [vij, vij]
             Temp = Max ({ f (Ai , cp , I) })ϵ cp;

          if Temp> MAXRMV then
           MAXRMV=Temp;
           else
          return vij as cut point for the cluster
```

5 Determining Reduct and Formation of Clusters

The data residing at each local site may have some irrelevant features, i.e. features which are having very negligible impact on the entire data set. These features are to be eliminated to reduce the complexity [12]. But the features discovered as irrelevant at one site may have considerable importance at other sites. Hence to derive the appropriate reduct set suitable for global modal the following procedure is used.

1. Reduct is found and clusters are formed at each local site.
2. The reduct set along with representatives of clusters formed are sent to the central site.
3. Once again reduct is found on the data set collected from various local modes and basing on the global reduct set clusters are formed.
4. The global reduct set along with cluster centroids are sent to local nodes, there clusters are modified according to the new centroids.

For finding reduct set the following algorithm is used.

Reduct(A)
A is set of all features

1 RED = {}
2 do
3 TEMP = RED
4 for each attribute x ϵ (A − RED)
5 for each attribute y ϵ A
6 $\gamma_{RED \cup \{x\}}(y) = \frac{|POS_{RED \cup \{X\}}(Y)|}{|U|}$
7 if $\overline{\gamma_{RED \cup \{x\}}(y), \forall y \epsilon A} > \overline{\gamma_{TEMP}(y), \forall y \epsilon A}$
8 TEMP ← RED ∪ {X}
9 RED ← TEMP
10 until $\overline{\gamma_{RED}(y), \forall y \epsilon A} = \overline{\gamma_{A}(y), \forall y \epsilon A}$
11 return RED

After determining the reduct and eliminate the reduct set from the data set. Clusters are formed. The data obtained from the images captured is unsupervised the next step is formation of clusters. The clustering techniques uses Rough set theory [13]. The technique makes use of roughness measurement and distance of relevance [14]. It works as follows:

Step 1: The relative roughness of each attribute with specific attribute value is calculated.
Relative Roughness: Relative Roughness of set X is measured basing on the lower and upper approximations of X, where X is a subset of objects having one specific value ∝ of attribute a_i with respect to $\{aj\}$ as

$$R_{aj}(X|ai = \propto) = 1 - \frac{\left|\underline{X_{aj}(ai = \propto)}\right|}{\left|\overline{X_{aj}(ai = \propto)}\right|} \quad \text{where } ai, aj \in \ and\ ai \neq aj$$

Step 2: Mean Roughness of each attribute is measured
Mean Roughness: The mean Roughness of the equivalence class $a_i = \propto$ denoted by $MeR(a_i = \propto)$ as

$$MeR(a_i = \propto) = \sum_{\substack{j=1 \\ j \neq i}}^{n} \frac{R_{aj}(X|ai = \propto)}{n - 1}$$

Step 3: After calculating the mean of each $a_i \in A$ we will apply standard deviation to each a_i by the formula

$$SD(ai = \propto) = \sqrt{1/(n - 1) \sum_{i=1}^{n-1} R_{aj}(X|ai = \propto) - MeR(a_i = \propto)}$$

Step 4: the value computed through standard deviation gives the splitting value to divide the table.

Step 5: After the table has been split into two Distance relevance of the two tables is calculated with formula

$$DR(B, C) = \sum_{i=1}^{n} (bi, ci)$$

Where *bi* is the value of object B and *ci* is that of C for ith attribute.

1. $DR(bi, ci) = 1$ if *bi* and *ci* are not equal
2. $DR(bi, ci) = 0$ otherwise

Step 5: Now distance Relevance of two tables is observed and whichever table has got highest DR is chosen for further splitting.

Step 6: Steps 1–5 are repeated until desired no of clusters are formed.

6 Transforming Clusters into Classes and Discovering Rules

The rules are discovered from the classes but the cosmetic data is unsupervised. Hence it is required to transform the unsupervised data into supervised data by attaching a class label to each cluster. The label attached is one of the attributes in the information table. The choice of attribute is done based on the dependency of each attribute on other attributes in the table. The dependency of the attribute is measured using Rough set theory approach.

The technique checks the maximum dependency of all the attributes in each cluster. Whichever attribute has highest maximum dependency that attribute is assigned as label to that cluster [15]. The procedure uses dependency of attributes in the rough set theory. First it computes the equivalence classes of each attribute. Then determines the dependency degree of each attribute using the formula

$$\gamma \mathrm{P(Q)} = \frac{\sum_{\mathrm{I=1}}^{\mathrm{N}} |\underline{\mathrm{P}}\mathrm{Q_I}|}{|\mathrm{U}|}$$

where $|\underline{\mathrm{P}}\mathrm{Q_I}|$ is the cardinality of lower approximation of equivalence class $\mathrm{Q_i}$ in $[\mathrm{X}]_\mathrm{Q}$.

Before we move to discovery of rules we go for feature extraction using the procedure described in Sect. 3. The rule extraction method generally uses discernibility matrix. The discernibility matrix suffers the problem of occupying large space. The storage for discernibility matrix can be reduced if we could eliminate all nonempty elements. Then rule extraction method is applied on this improved discernibility matrix [16].

Given a decision table S = (U, C, D, V, f), let $M = (m_{ij})_{n \times n}$ be discernibility matrix whose elements are defined as follows:

$$m_{ij} = \begin{cases} \{c_k \in C, f(x_i, c_k) \neq f(x_j, c_k) \bigwedge f(x_i, D) \neq f(J, D)\} \\ \emptyset \quad else \end{cases}$$

Given a decision table S = (U, C, D, V, f), for $v \in V_c, w \in V_d$, the form of decision rules, $\bigwedge(c_i, v) \rightarrow \bigvee(d, w)$, where $\bigwedge(c_i, v)$ is condition element of rules and $\bigvee(d, w)$ s decision element of rules.

Algorithm Efficient rule extraction algorithm based on discernibility matrix in decision table

Input: Decision table $S = (U, C, D, V, f)$, where domain U = $\{x_1, x_2, \ldots, x_\mathrm{n}\}$, conditional attributes $C = \{c_1, c_2, \ldots, c_\mathrm{m}\}$, decision attribute $D = \{d\}$.
Output: Rule sets *Rule* (*U*).
Begin:

Step 1: discernibility matrix $M' = \emptyset$; knowledge reduction $Red = \emptyset$;
Step 2: computing equivalences classes i.e. partition:$U/D = \{D_1, D_2, D_3 \ldots\}$
Step 3: for $\forall x_i, x_j \in U and f(x_i, d) \neq f(x_{ij}, d)$, according to definition of discernibility matrix for $\forall x c_{jk} \in C$ if $f(x, c_k) \neq f(y, c_k)$ then $m(x_i, x_j) = m(x_i, x_j) \cap \{\mathrm{c_k}\}$

Step 4: for each non-empty element $m(x_i, x_j)$ in discernibility matrix M'
Step 4.1: if $m(x_i, x_j)$.count $==$ 1 then $RedRed \cup (c_i)$
Step 4.2: for each non-empty element $m(x_i, x_j)$, if $c_i \in m(x_i, x_j)$, then delete matrix element $m(x_i, x_j)$ in discernibility matrix M
Step 5: for each attribute c_i in discernibility matrix M, compute the count of each c_i
Step 5.1: Select an attribute satisfying the maximal count c_k
If there exists the attribute c_k,then $Red = Red \cup \{i\}$
Step 5.2: for each non-empty element (x_i, x_j), if $c_i \in m(x_i, x_j)$,then delete matrix element $m(x_i, x_j)$ in discernibility matrix M
Step 6: if discernibility matrix M is not empty, then turn to Step 5
else output knowledge reduction Red
Step 7: according to the knowledge reduction Red, general rule sets is obtained
Step 8: output the simplest rule sets Rule (U)

End

The rules generated are sent to the central node where regeneration is done by eliminating the repetitive components and final rules are generated.

7 Results

The experimental results are presented in this section. The entire process is applied on sample cosmetic data which consists of 18 features. To show the experimental results a dataset of 15 samples at each local site is taken which consists of 17 numeric

Decision rules
(Stype=1)&(Saa=24.6)&(s_count=5)=>(skin_problem={pimple[1]})
(Stype=5)&(Saa=4.5)&(s_count=88)=>(skin_problem={e_wrinkles[1]})
(Stype=6)&(Saa=0.3)&(s_count=41)=>(skin_problem={e_wrinkles[1]})
(Stype=5)&(Saa=10.3)&(s_count=91)=>(skin_problem={e_wrinkles[1]})
(Stype=5)&(Saa=3.4)&(s_count=41)=>(skin_problem={e_wrinkles[1]})
(Stype=5)&(Saa=3.8)&(s_count=78)=>(skin_problem={e_wrinkles[1]})
(Stype=5)&(Saa=2.6)&(s_count=53)=>(skin_problem={e_wrinkles[1]})
(Stype=6)&(Saa=2.4)&(s_count=119)=>(skin_problem={e_wrinkles[1]})
(Stype=5)&(Saa=0.4)&(s_count=89)=>(skin_problem={e_wrinkles[1]})
(Stype=5)&(Saa=2.29)&(s_count=146)=>(skin_problem={e_wrinkles[1]})
(Stype=5)&(Saa=3.1)&(s_count=60)=>(skin_problem={e_wrinkles[1]})
(Stype=1)&(Saa=30.9)&(s_count=21)=>(skin_problem={pimple[1]})
(Stype=1)&(Saa=48.9)&(s_count=83)=>(skin_problem={pimple[1]})
(Stype=5)&(Saa=19.3)&(s_count=169)=>(skin_problem={cysts[1]})
(Stype=5)&(Saa=65.5)&(s_count=315)=>(skin_problem={cysts[1]})
(Stype=5)&(Saa=3.8)&(s_count=176)=>(skin_problem={cysts[1],cluster0[1]})

Fig. 2 Rules discovered

Table 1 Features extracted from face images of customers in cosmetic industry sample data set

Age	Skin type	Spot count	Age spot	Pimples	Pastules	Papules	Area affected	Wrinkles	Hydration	Cysts	Break out visibility	Acne count	Pore count	Visible pores	Emerging lines	Fine Lines	Deep lines
22	5	59	3	4	0	0	1.5	0	63.2	0	3.96	12	1204	2.8717	44	16	5
23	5	52	2	6	0	0	14.3	0	77.2	0	1.55	8	1597	3.0032	32	9	6
25	6	49	1	5	0	1	0.8	0	80.4	0	1.89	8	1200	2.5261	57	10	1
22	5	21	1	0	0	0	1.1	0	75.4	0	0	10	1351	2.5792	42	7	2
22	5	16	1	0	0	0	5.7	0	71.7	0	0	1	867	1.8601	18	4	5
25	5	44	2	3	0	0	2.6	0	74.9	0	1.53	10	1197	2.6867	33	9	0
20	5	44	2	2	0	0	6.2	0	75.3	0	1.371	9	1225	2.5490	19	4	3
25	5	115	5	9	0	1	3.8	0	55.3	0	2.56	22	1512	3.1148	72	8	0
23	4	21	1	0	0	0	3.9	0	77.3	0	0	1	427	1.0588	38	11	3
29	1	5	1	0	0	0	24.6	0	84.7	0	0	0	746	1.4453	11	1	0
29	5	88	3	2	0	0	4.5	0	66.3	0	4.2	23	1468	3.1759	80	16	1
33	6	41	2	2	0	0	0.3	0	78.4	0	2.597	4	1372	3.4534	67	14	2
32	5	91	6	3	0	0	10.3	0	61.1	0	5.70	5	1263	2.6082	30	13	41
33	5	41	3	0	0	0	3.4	0	61.6	0	0	4	1135	2.3677	29	3	3
29	5	78	4	6	0	1	3.8	0	72.3	0	3.21	24	1170	2.546218	37	12	11
35	5	53	2	0	0	0	2.6	0	58.3	0	0	6	1140	2.3873	27	3	0
35	6	119	10	15	0	0	2.4	0	55.1	1	3.10	20	984	4.5490	99	57	18
35	5	89	4	1	0	0	0.4	0	65	0	1.53	24	1881	5.9193	81	23	12

Table 2 No of intervals formed after discretization

S no	Attribute	Type	Distinct values (distict values)	Intervals
1	S type	Numeric	5	3
2	Saa	Numeric	30	7
3	S_Count	Numeric	29	6
4	A_Spots	Numeric	12	4
5	Pimples	Numeric	13	4
6	Pastules	Numeric	3	3
7	Papules	Numeric	5	4
8	Cysts	Numeric	2	2
9	B_Visi	Numeric	23	4
10	A_count	Numeric	20	4
11	p_count	Numeric	33	8
12	V_pores	Numeric	33	4
13	E_lines	Numeric	31	6
14	F_Lines	Numeric	27	7
15	D_lines	Numeric	20	7
16	E_Wrinkles	Numeric	1	1
17	h_skin	Numeric	33	5

Table 3 Generated reduct

Iteration	Reduct
1	Saa
2	Saa, E_lines
3	Saa, E_lines, F_Lines
4	Saa, E_lines, F_Lines, pimples
5	Saa, E_lines, F_Lines, pimples, Stype
6	Saa, E_lines, F_Lines, pimples, Stype, e_wrinkles
7	Saa, E_lines, F_Lines, pimples, Stype, e_wrinkles, pastules
8	Saa, E_lines, F_Lines, pimples, Stype, e_wrinkles, pastules, S_Count
9	Saa, E_lines, F_Lines, pimples, Stype, e_wrinkles, pastules, S_Count, B_Visi
10	Saa, E_lines, F_Lines, pimples, Stype, e_wrinkles, pastules, S_Count, B_Visi, A_count

Features. After applying the proposed algorithms the discovered rules at the local site are given in Fig. 2 (Tables 1, 2, 3).

As the industry is so far using the manual trial and error based method the proposed method gives results with adequate accuracy and thus a time saving procedure.

8 Conclusion

Rough set theory concepts are used in analyzing and discovering rules in cosmetic data in distributed environment. The analysis can be done either at the central node collecting information from each site or at each local site collecting knowledge from other parallel sites without the need of the central site. The later one reduces the complexity as central repository need not be maintained.

References

1. Cox NH, Coulson IH (2004) Diagnosis of skin disease. vol 1, 7th edn. Department of Dermatology, Cumberland Infirmary, Carlisle, UK Dermatology Unit, Burnley General Hospital, Burnley, UK
2. Barati E, Saraee M, Mohammadi A, Adibi N, Ahamadzadeh MR (2011) A survey on utilization of data mining approaches for dermatological (skin) diseases prediction cyber journals: multidisciplinary journals in science and technology. J Sel Areas Health Informatics (JSHI)
3. Komorowski J, Polkowski L, Skowron A (2001) Rough sets: a tutorial. Fundamental Informaticae XXI:1001–1025 (1001 IOS Press)
4. Pawlak Z (03/2002) Rough set theory and its applications. J Telecommun inf Technol
5. Skowron A, Peters JF (2003) Rough sets: trends and challenges. Extended abstract, rough sets, fuzzy sets, data mining, and granular computing. Lect Notes Comput Sci 2639:25–342
6. Suraj Z (2004) An introduction to rough set theory and its applications. ICENCO'2004, 27–30 Dec, Cairo, Egypt
7. Walczak B, Massart DL (Feb 2009) Rough set theory—fundamental concepts, principals, data extraction, and applications. In: Ponce J, Karahoca A (eds) Data mining and knowledge discovery in real life applications. I-Tech, Vienna, Austria, pp 438. ISBN 978-3-902613-53-0
8. Kotsiantis S, Kanellopoulos D (2006) Discretization techniques: a recent survey. GESTS Int Trans Comput Sci Eng 32(1):47–58
9. Singh GK, Minz S (2007) Discretization using clustering and rough set theory. In: Proceedings of the international conference on computing: theory and applications (ICCTA'07) 0-7695-2770-1/07
10. Sengupta N (May 2012) Evaluation of rough set theory based network traffic data classifier using different discretization method. Int J Inf Electron Eng 2(3)
11. Kanungo T, Mount DM, Netanyahu NS, Piatko CD, Silverman R, Wu AY (July 2002) An efficient k-means clustering algorithm: analysis and implementation. IEEE Trans Pattern Anal Mach Intell 24(7)
12. Dash M, Liu H (2000) Unsupervised feature selection. In: Proceedings of the pacific and asia conference on knowledge discovery and data mining, Kyoto, pp 110–121
13. Tripathy BK, Ghosh A (2011) SSDR: an algorithm for clustering categorical data using rough set theory. Pelagia Res Libr Adv Appl Sci Res 2
14. Han J, Sanchez R, Hu X, Lin TY (2005) Feature selection based on relative attribute dependency: an experimental study. Department of Computer Science, California State University Dominguez Hills 1000. In: The tenth international conference on rough sets, fuzzy sets, data mining, and granular computing (RSFDGrC 2005) Aug 31st–Sept 3rd, University of Regina, Canada
15. Herawan T, Deris MM, Abawajy JH (2010) A rough set approach for selecting clustering attribute. Knowl Based Syst 23(3):220–223 (University of Tun Hussein Onn Malaysia, Faculty of Information Technology and Multimedia)
16. Zou H, Zhang C (April 2012) Efficient rule extraction algorithm based on discernibility matrix in decision table. Int J Advancements Comput Technol (IJACT) ROU 4(6). doi:10.4156/ijact.vol4.issue6.35

Printed in the United States
By Bookmasters